"十二五"江苏省高等学校重点教材

编号：2014-1-119

数据库原理及应用实验教程

第二版

主　编　吴克力　陈　雅

主　审　李存华

南京大学出版社

图书在版编目(CIP)数据

数据库原理及应用实验教程 / 吴克力,陈雅主编.
—2 版. — 南京:南京大学出版社,2016.6
 21 世纪应用型本科计算机专业实验系列教材
 ISBN 978-7-305-17261-8

Ⅰ. ①数… Ⅱ. ①吴… ②陈 Ⅲ. ①关系数据库系统—高等学校—教材 Ⅳ. ①TP311.138

中国版本图书馆 CIP 数据核字(2016)第 152588 号

出版发行	南京大学出版社
社　　址	南京市汉口路 22 号　　邮　编　210093
出 版 人	金鑫荣
丛 书 名	21 世纪应用型本科计算机专业实验系列教材
书　　名	数据库原理及应用实验教程
主　　编	吴克力　陈　雅
责任编辑	王秉华　　　　　编辑热线　025-83595860
照　　排	南京理工大学资产经营有限公司
印　　刷	南京京新印刷厂
开　　本	787×960　1/16　印张 16.25　字数 345 千
版　　次	2016 年 6 月第 2 版　2016 年 6 月第 1 次印刷
ISBN	978-7-305-17261-8
定　　价	35.00 元

网　　址:http://www.njupco.com
官方微博:http://weibo.com/njupco
官方微信号:njupress
销售咨询热线:(025)83594756

＊版权所有,侵权必究
＊凡购买南大版图书,如有印装质量问题,请与所购
　图书销售部门联系调换

第二版前言

《数据库原理及应用》是四年制本科计算机及相关专业的一门重要必修课。该课程的理论性、实践性和应用性均很强。教学内容既有抽象的关系数据库基础理论，又有实践性极强的数据库设计和应用软件开发。理论与实验教学之间相辅相成，缺一不可。实验教学能化抽象为直观，加深基础理论的理解，增强学生的实践动手能力。本书以流行的微软公司数据库管理系统 SQL Server 2008 和软件设计工具 Visual Studio 2010 为平台，程序设计采用 C♯语言。

本书配合数据库原理及应用课程的实验教学编写，理论教材建议采用王珊、萨师煊编著的《数据库系统概论》（第四版）。实验教程的内容与理论教学章节紧密衔接，依次为 Windows 窗体应用程序设计基础、SQL 语句实验、存储过程与管理系统使用、报表服务器应用、ADO.NET 应用软件开发技术、学生成绩管理系统和课程设计案例与选题。

全书共分 6 章。第 1 章为 Visual Studio 2010 C♯程序设计基础，通过鼠标点击坐标、捉迷藏的按钮、移动的窗体、登录窗体、省市选择器、树型组织机构、电子表、验证码图像的生成、MP3 播放器、正则表达式等实验示例，学习窗体应用程序的设计，为后续章的学习和课程设计奠定基础。

第 2 章为 SQL Server 2008 数据库应用基础，主要结合理论教学中数据库创建、SQL 语句、数据库完整性控制、存储过程、安全性控制、数据备份与恢复等知识设计了多组实验，内容涵盖了理论教材中重要的 SQL 基础语句和数据库管理系统的使用方法。

第 3 章为数据库应用程序设计基础，介绍了基于 ADO.NET 进行数据库应用软件开发的常用技术，包括数据库连接与维护、主从数据表、LINQ 技术、图片在数据库中的存储、水晶报表、服务器报表和安装程序制作等。

第 4 章以学生成绩管理系统的设计为例，介绍了数据库应用软件所具有的主要功能、数据库设计方法和编程方法。

第 5 章根据数据库课程设计的要求，给出了一个较为完整的课程设计报告，供学习者撰写课程设计报告时参考。

第 6 章主要是课程设计选题，供分派课程设计任务时参考。

本书第 1 章由吴克力和陈雅共同编写，第 2、3 章由吴克力编写，第 4、5、6 章由陈雅编写，全书由吴克力完成统稿，淮海工学院李存华教授认真审阅了全部书稿，并提出了十分珍贵的修改意见，在此表示最诚挚的谢意！

本书第一版为江苏省高等学校精品教材,第二版被遴选为"十二五"江苏省高等学校重点教材,根据读者反馈和教学实践,第二版从以下几个方面进行了改进:① 第 1 章增强了应用程序设计基础实验,删除了部分设计难度较大的实验示例,使教材更适合初学者。② 基于新的软件开发平台,引入了数据库报表服务器布置、应用开发等内容,用功能更加强大的服务器报表代替水晶报表,考虑到向上兼容保留了水晶报表设计章节,此外,删去了应用较少的数据库邮件部分内容。③ 第 5 章用一个完整的课程设计替换原教材中设计难度较大的基于 Web 的短信群发系统。④ 本书大部分实验项目配备有电子资源,读者可以通过手机扫二维码的形式进行学习,体现了数字出版和教材立体化建设的理念。

在本书编写过程中,还参考了国内外数据库及程序设计等相关书籍,在此表示谢意。本书可作为大学本科、高职高专院校的数据库实验教材,也可作为初学者学习数据库的参考。

由于作者水平所限,错误和不足之处在所难免,敬请读者和专家批评指正。读者的建议、意见可以通过发电子邮件到 wkl@hytc.edu.cn 或 juliachenya@163.com 与我们联系。

编 者

2016 年 4 月

目 录

第1章 Visual Studio 2010 C#程序设计基础 ·········· 1

1.1 窗体应用程序设计基础 ·········· 1
1.1.1 鼠标点击坐标示例程序 ·········· 1
1.1.2 捉迷藏的按钮示例程序 ·········· 3
1.1.3 移动的窗体示例程序 ·········· 5
1.1.4 登录窗体示例程序 ·········· 8
1.1.5 省市选择器示例程序 ·········· 10
1.1.6 树型组织机构示例程序 ·········· 15

1.2 图形图像程序设计基础 ·········· 18
1.2.1 电子表示例程序 ·········· 18
1.2.2 线的绘制示例程序 ·········· 21
1.2.3 验证码图像的生成示例程序 ·········· 25

1.3 MP3播放器示例程序 ·········· 29
1.4 文件操作示例程序 ·········· 35
1.5 正则表达式示例程序 ·········· 38

第2章 SQL Server 2008 数据库应用基础 ·········· 41

2.1 数据库创建与管理 ·········· 41
2.1.1 数据库的创建与删除 ·········· 41
2.1.2 认识SQL Server数据库 ·········· 43

2.2 表的操作与视图 ·········· 44
2.2.1 创建表 ·········· 44
2.2.2 索引 ·········· 46
2.2.3 输入数据 ·········· 47
2.2.4 数据查询 ·········· 50
2.2.5 关系图 ·········· 55

  2.2.6　视图 …………………………………………………………………………… 57
2.3　数据库安全性 ………………………………………………………………………… 58
  2.3.1　服务器身份验证模式 ………………………………………………………… 58
  2.3.2　登录名与服务器角色 ………………………………………………………… 59
  2.3.3　用户与数据库角色 …………………………………………………………… 61
2.4　存储过程与触发器 …………………………………………………………………… 64
  2.4.1　存储过程 ……………………………………………………………………… 64
  2.4.2　触发器 ………………………………………………………………………… 65
2.5　数据库的维护 ………………………………………………………………………… 67
  2.5.1　分离和附加数据库 …………………………………………………………… 67
  2.5.2　脱机和联机数据库 …………………………………………………………… 68
  2.5.3　备份与还原数据库 …………………………………………………………… 68
  2.5.4　导入与导出数据库 …………………………………………………………… 70
  2.5.5　制定维护计划 ………………………………………………………………… 72
2.6　Reporting Services 应用 ……………………………………………………………… 73
  2.6.1　Reporting Services 配置管理器 ……………………………………………… 73
  2.6.2　报表设计 ……………………………………………………………………… 75
  2.6.3　报表部署 ……………………………………………………………………… 77

第 3 章　数据库应用程序设计基础 …………………………………………………………… 78

3.1　应用程序与数据源连接的建立 ……………………………………………………… 78
  3.1.1　ADO.NET 简介 ………………………………………………………………… 78
  3.1.2　连接数据源 …………………………………………………………………… 79
  3.1.3　用配置文件保存连接字符串 ………………………………………………… 80
  3.1.4　连接数据源示例程序 ………………………………………………………… 81
3.2　用控件显示数据 ……………………………………………………………………… 85
  3.2.1　DataSet、DataTable 和 DataAdapter 对象 …………………………………… 85
  3.2.2　表中数据读至 DataSet 对象 ………………………………………………… 86
  3.2.3　数据显示控件与数据表的绑定 ……………………………………………… 87
  3.2.4　用控件显示数据示例程序 …………………………………………………… 87
3.3　数据的插入、修改与删除 …………………………………………………………… 90
  3.3.1　用 ADO.NET 维护数据表 …………………………………………………… 90
  3.3.2　数据表维护示例程序 ………………………………………………………… 91

目录

- 3.4 调用数据库存储过程 ……………………………………………… 96
  - 3.4.1 ADO.NET 调用存储过程 ……………………………………… 96
  - 3.4.2 调用存储过程示例程序 ……………………………………… 97
- 3.5 数据库中图像的存取 ………………………………………………… 100
  - 3.5.1 图像存储的数据类型 ………………………………………… 100
  - 3.5.2 图像存取方法 ………………………………………………… 101
  - 3.5.3 数据库中图像存取示例程序 ………………………………… 101
- 3.6 主从关系数据表 ……………………………………………………… 109
  - 3.6.1 主从数据表 …………………………………………………… 109
  - 3.6.2 主从数据表示例程序 ………………………………………… 109
- 3.7 语言集成查询(LINQ)技术 ………………………………………… 111
  - 3.7.1 LINQ 简介 …………………………………………………… 111
  - 3.7.2 对象关系设计器 ……………………………………………… 112
  - 3.7.3 查询表达式 …………………………………………………… 113
  - 3.7.4 LINQ to SQL 访问 SQL 数据库 …………………………… 115
  - 3.7.5 LINQ 技术访问数据库示例程序 …………………………… 115
- 3.8 服务器报表应用程序设计 …………………………………………… 121
  - 3.8.1 ReportViewer 控件 …………………………………………… 121
  - 3.8.2 服务器报表示例程序 ………………………………………… 121
- 3.9 水晶报表应用程序设计 ……………………………………………… 124
  - 3.9.1 水晶报表基础 ………………………………………………… 124
  - 3.9.2 水晶报表设计器 ……………………………………………… 125
  - 3.9.3 报表数据源 …………………………………………………… 126
  - 3.9.4 水晶报表示例程序 …………………………………………… 126
- 3.10 安装程序制作 ……………………………………………………… 130
  - 3.10.1 Microsoft Windows Installer 程序简介 …………………… 130
  - 3.10.2 创建安装程序 ……………………………………………… 130
  - 3.10.3 创建卸载程序 ……………………………………………… 131
  - 3.10.4 安装程序的设计示例程序 ………………………………… 131

第 4 章 学生成绩管理系统 …………………………………………… 134

- 4.1 系统设计 ……………………………………………………………… 134
  - 4.1.1 系统需求及功能概述 ………………………………………… 134

  4.1.2 概念结构设计 ·············· 135
  4.1.3 数据库设计 ·············· 135
 4.2 系统详细设计 ·············· 137
  4.2.1 登录界面设计 ·············· 137
  4.2.2 管理员主界面设计 ·············· 140
  4.2.3 系统管理 ·············· 142
  4.2.4 学生管理 ·············· 146
  4.2.5 课程管理、教师管理、成绩管理 ·············· 153
  4.2.6 成绩统计 ·············· 153
  4.2.7 教师身份主界面设计 ·············· 158

第5章 课程设计案例 ·············· 164

 5.1 数据库应用软件设计步骤 ·············· 164
 5.2 需求分析 ·············· 164
  5.2.1 项目需求 ·············· 165
  5.2.2 系统功能需求 ·············· 165
  5.2.3 安全性、完整性要求 ·············· 166
 5.3 概念结构设计 ·············· 166
 5.4 数据库逻辑结构设计 ·············· 167
  5.4.1 关系模式设计 ·············· 167
  5.4.2 子模式设计 ·············· 172
 5.5 系统详细设计 ·············· 172
  5.5.1 配置文件设置 ·············· 172
  5.5.2 登录界面设计 ·············· 173
  5.5.3 主界面设计 ·············· 179
  5.5.4 系统管理 ·············· 186
  5.5.5 宿舍信息 ·············· 192
  5.5.6 学生入住 ·············· 200
  5.5.7 卫生检查 ·············· 210
  5.5.8 水电收费 ·············· 217
  5.5.9 房屋报修 ·············· 223
  5.5.10 外来人员登记 ·············· 227
  5.5.11 宿舍入住统计报表 ·············· 230

第6章 数据库课程设计选题 ····· 235

6.1 课程设计基本要求 ····· 235
6.2 课程设计选题 ····· 236
  6.2.1 工资管理系统 ····· 236
  6.2.2 教务管理系统 ····· 237
  6.2.3 图书管理系统 ····· 238
  6.2.4 客房管理系统 ····· 240
  6.2.5 民航订票管理系统 ····· 242
  6.2.6 学生信息管理系统 ····· 243
  6.2.7 长途汽车信息管理系统 ····· 244
  6.2.8 其他参考选题 ····· 245
6.3 课程设计报告要求 ····· 246

参考文献 ····· 249

第1章 Visual Studio 2010 C♯程序设计基础

Visual Studio 2010 是微软公司推出的集成化软件开发环境,它继承了先前版本使用简单、功能强大、效率高等特点,已成为在 Windows 环境下开发应用程序的首选工具。其可视化的快速应用程序开发(RAD)能力,能帮助开发人员快速设计 Windows 窗体应用程序、Web 应用程序、WPF 应用程序以及移动应用程序。

C♯是微软公司在. NET Framework 平台上首推的程序开发语言。它既有 PASCAL 语言语法严谨的特点,又有 C++、JAVA 等现代程序设计语言所具有的面向对象的强大功能,是一种简单、现代、通用、面向对象的编程语言。

本章通过若干个示例程序的设计,学习 Visual Studio 2010 设计应用程序的基本方法。

1.1 窗体应用程序设计基础

1.1.1 鼠标点击坐标示例程序

Visual Studio 2010 具有一个非常易用、功能强大的集成化开发环境,可以生成、测试和调试 C♯应用程序。开发环境的默认布局如图 1-1,最上端是菜单栏和工具栏,左侧是"工具箱"窗口。右上是显示项目所有文件视图的"解决方案资源管理器"窗口。右下是控件的"属性"窗口。中间有起始页页面、用于可视化设计的"窗体设计器"窗口页面、用于编写和调试程序的"代码编辑器"窗口页面。

在 VS 2010 开发环境中,应用程序的界面设计为所见即所得,用户从工具箱中拖放按钮、文本框、下拉列表等控件到设计窗口,再根据需求设置控件属性,编写控件的事件响应函数,反复调试修改,即可完成应用程序的开发。

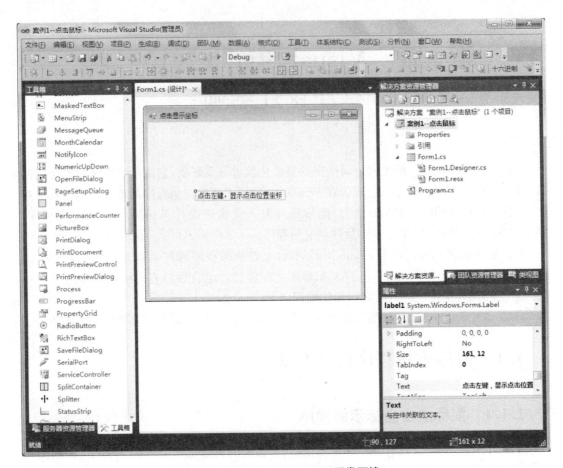

图 1-1　Visual Studio 2010 开发环境

【实验 1-1】　设计一个显示鼠标点击信息的程序。当鼠标左键点击窗体，在窗体的点击处显示坐标值。

相关控件：

属性与响应事件见表 1-1。

表 1-1　实验 1-1 控件设置与响应事件

控件	名称	属性值	事件
Form	Form1	Text=点击显示坐标	MouseClick
Label	Label1	Text=点击左键,显示点击位置坐标	

主要代码:

```
private void Form1_MouseClick(object sender, MouseEventArgs e)
{
    label1.Text = "(" + e.X.ToString() + "," + e.Y.ToString() + ")";
    label1.Location = new Point(e.X, e.Y);
}
```

关键技术:

(1) Windows 系统的应用程序是事件驱动程序,通常点击鼠标、键盘按键等输入动作将触发事件的发生,事件发生将引起对应的代码被执行。本例就是当鼠标左键点击程序窗体时,触发了窗体的 MouseClick 鼠标事件,从而引起修改 label1 控件的代码被执行,实现了 Label 控件的 Text 内容为鼠标坐标,并且控件的左上角位置被改为鼠标点击位置。

(2) 鼠标能触发的事件较多,常用的有 MouseClick,MouseDoubleClick,MouseDown,MouseUp,MouseEnter,MouseHover,MouseLeave,MouseMove 等。

【实验项目】

1.1 在实验 1-1 程序中添加下面功能:(1) 点击鼠标右键,显示"你点了鼠标左键!"字符串;(2) 鼠标进入文字显示区域,文字的背景变红,离开则恢复正常。

1.2 设计一个点击按钮(或超链接),调用 IE 访问指定网站的程序。

1.3 编程熟悉 Button,CheckBox,CheckedListBox,ComboBox,DateTimePicker,ListBox,TextBox,RadioButton 等控件的功能。

1.1.2 捉迷藏的按钮示例程序

【实验 1-2】 设计一个按钮会捉迷藏的程序,当你点击按钮时,按钮会躲避点击,运行效果如图 1-2 所示。

相关控件:

属性设置与响应事件见表 1-2。

表1-2 实验1-2控件属性及响应事件

控件	名称	属性值	事件
Form	Form1	Text＝捉迷藏的按钮	
Button	BtnHide	Text＝你点我呀	MouseEnter

图1-2 捉迷藏的按钮

主要代码：

```
private void BtnHide_MouseEnter(object sender, EventArgs e)
{
    Random ran = new Random();
    int x = ran.Next(0, this.ClientSize.Width - BtnHide.Width);
    int y = ran.Next(0, this.ClientSize.Height - BtnHide.Height);
    BtnHide.Location = new Point(x, y);
}
```

关键技术：

（1）由 Button 的 MouseEnter 事件触发 Button 的位置改变。

（2）注意窗体的 ClientSize 与 Size 属性的区别。

（3）要防止 Button 随机移动位置时跳出窗体活动区域，思考随机坐标的取值范围。

【实验项目】

1.4 设计一个如图1-3所示窗体，点击"喜欢"按钮，弹出消息框，显示"我也喜欢你"，若点击"不喜欢"按钮，则按钮随机移动鼠标无法点中。

第1章 Visual Studio 2010 C#程序设计基础

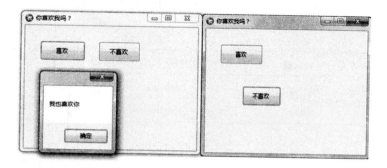

图 1-3 你喜欢我吗?

1.1.3 移动的窗体示例程序

【实验1-3】 设计会移动的窗体程序,点击相应的按钮,窗体作相应方向的移动,运行效果如图1-4所示。

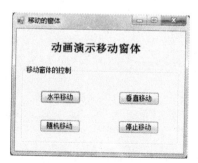

图 1-4 移动的窗体

相关控件:

各控件相关属性及事件如表1-3所示。

表 1-3 控件属性设置及事件介绍

控件	名称	属性值	事件
Form	Form1	Text=移动的窗体	
Button	BtnX	Text=水平移动	Click
	BtnY	Text=垂直移动	Click
	BtnRandom	Text=随机移动	Click
	BtnStop	Text=随机移动	Click

(续表)

控件	名称	属性值	事件
Timer	timer1	Enabled=false InterVal=100	Tick
	timer2	Enabled=false InterVal=100	Tick
	timer3	Enabled=false InterVal=2000	Tick

主要代码：

（1）定义外部变量：

```csharp
int ScreenWidth = SystemInformation.PrimaryMonitorMaximizedWindowSize.Width;
int ScreenHeight = SystemInformation.PrimaryMonitorMaximizedWindowSize.Height;
```

（2）事件代码：

```csharp
private void timer1_Tick(object sender, EventArgs e)
{
    Point mypos = new Point(DesktopLocation.X, DesktopLocation.Y);
    if (mypos.X + Width < ScreenWidth)
    {
        this.DesktopLocation = new Point(mypos.X + 2, mypos.Y);
    }
    else
    {
        DesktopLocation = new Point(0, 0);
    }
}

private void timer2_Tick(object sender, EventArgs e)
{
    Point mypos=new Point(DesktopLocation.X,DesktopLocation.Y);
    if(mypos.Y+Height<ScreenHeight)
    {
        DesktopLocation=new Point(mypos.X,mypos.Y+2);
    }
    else
    {
        DesktopLocation=new Point(0,0);
    }
}
```

```
    }

    private void timer3_Tick(object sender, EventArgs e)
    {
        Random ran = new Random();
        int x = ran.Next(1, this.ClientSize.Width);
        int y = ran.Next(1, ScreenHeight);
        Point mypos = new Point(DesktopLocation.X, DesktopLocation.Y);
        DesktopLocation = new Point(x, y);
    }

    private void BtnX_Click(object sender, EventArgs e)
    {
        timer1.Enabled = true;
        timer2.Enabled = false;
        timer3.Enabled = false;
    }

    private void BtnY_Click(object sender, EventArgs e)
    {
        timer1.Enabled = false;
        timer2.Enabled = true;
        timer3.Enabled = false;
    }

    private void BtnRandom_Click(object sender, EventArgs e)
    {
        timer1.Enabled = false;
        timer2.Enabled = false;
        timer3.Enabled = true;
    }

    private void BtnStop_Click(object sender, EventArgs e)
    {
        timer1.Enabled = false;
```

```
        timer2.Enabled = false;
        timer3.Enabled = false;
}
```

关键技术：

(1) 通过引发 Timer_Tick 事件，Timer 控件可以有规律地隔一段时间执行一次代码；

(2) 掌握 Timer 控件的 Enabled 属性、Interval 属性的含义；

(3) 掌握 Timer 控件的 Tick 事件在何时被触发。

1.1.4 登录窗体示例程序

【实验 1-4】 设计登录窗体，输入正确的用户名和密码后，点击登录按钮可以登录进入主窗体，否则，弹出对话框显示错误，运行效果如图 1-5 所示。

相关控件：

各控件相关属性及事件如表 1-4 所示。

表 1-4 控件属性设置及事件介绍

控件		名称	属性值	事件
Form		Form1	Text=登录窗体	
			TopMost=true	
			StartPosition=CenterScreen	
		Form2	Text=主窗体	
Form1	Button	BtnLogin	Text=登录	Click
		BtnCancel	Text=取消	Click
		BtnExit	Text=退出	Click
	Label	label1	Text=登录系统	
		label2	Text=用户名	
		label3	Text=密码	
	TextBox	txtusername		
		txtpassword	PasswordChar=*	
Form2	Label	label1	Text=恭喜您,登录成功	

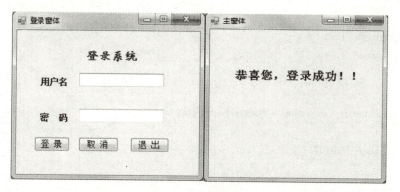

图 1-5 登录窗体

主要代码:

```
private void btnlogin_Click(object sender, EventArgs e)
{
    if (txtusername.Text == "" || txtpassword.Text == "")
        MessageBox.Show("登录信息不完整","友情提醒",MessageBoxButtons.OK, MessageBoxIcon.Warning);
    else
    {
        if (txtusername.Text == "悠悠" && txtpassword.Text == "888888")
        {
            Form2 f2 = new Form2();
            f2.Show();
            this.Hide();
        }
        else
        {
            MessageBox.Show("用户名或密码不正确,请重新输入","友情提醒",MessageBoxButtons.OK, MessageBoxIcon.Information);
            txtusername.Focus();
            txtusername.SelectAll();
        }
    }
}

private void btncancel_Click(object sender, EventArgs e)
```

```
{
    txtusername.Text = "";
    txtpassword.Text = "";
}

private void btnexit_Click(object sender, EventArgs e)
{
    Application.Exit();
}
```

关键技术：
(1) 通过设置窗体的 TopMost 属性设置窗体始终在最前；
(2) 掌握用户名、密码匹配的判断方法；
(3) 登录成功后，跳转至其他窗体的实现方法。

【实验项目】
1.5 在示例 1-4 基础上，登录成功后，在主窗体的 Label 中，显示"恭喜您"＋登名＋"成功登录"。

1.1.5 省市选择器示例程序

【实验 1-5】 设计省市联动选择程序，通过 ComboBox 控件实现，运行效果如图 1-6 所示。

相关控件：
各控件相关属性及事件如表 1-5 所示。

表 1-5 控件属性设置及事件介绍

控件	名称	属性值	事件
Form	Form1	Text=省市选择器	Load
Button	btnshow	Text=显示	Click
Form1.Label	label1	Text=省份	
	label2	Text=城市	
	lblshow	Text=""	

（续表）

控件	名称	属性值	事件
ComboBox	Cboprovince		SelectedIndexChanged
	CboCity		

图 1-6　省市选择器

主要代码（仅以山西、河北、内蒙古为例）：

```
private void Form1_Load(object sender, EventArgs e)
{
    CboCity.DropDownStyle = ComboBoxStyle.DropDownList;
    Cboprovince.DropDownStyle = ComboBoxStyle.DropDownList;
    Cboprovince.Items.Add("北京");
    Cboprovince.Items.Add("上海");
    Cboprovince.Items.Add("天津");
    Cboprovince.Items.Add("重庆");
    Cboprovince.Items.Add("河北");
    Cboprovince.Items.Add("山西");
    Cboprovince.Items.Add("内蒙古");
    Cboprovince.Items.Add("辽宁");
    Cboprovince.Items.Add("吉林");
    Cboprovince.Items.Add("黑龙江");
    Cboprovince.Items.Add("江苏");
    Cboprovince.Items.Add("浙江");
    Cboprovince.Items.Add("安徽");
    Cboprovince.Items.Add("福建");
```

```
            Cboprovince.Items.Add("江西");
            Cboprovince.Items.Add("山东");
            Cboprovince.Items.Add("河南");
            Cboprovince.Items.Add("湖北");
            Cboprovince.Items.Add("湖南");
            Cboprovince.Items.Add("广东");
            Cboprovince.Items.Add("广西");
            Cboprovince.Items.Add("海南");
            Cboprovince.Items.Add("四川");
            Cboprovince.Items.Add("贵州");
            Cboprovince.Items.Add("云南");
            Cboprovince.Items.Add("西藏");
            Cboprovince.Items.Add("陕西");
            Cboprovince.Items.Add("甘肃");
            Cboprovince.Items.Add("青海");
            Cboprovince.Items.Add("宁夏");
            Cboprovince.Items.Add("新疆");
            Cboprovince.Items.Add("香港");
            Cboprovince.Items.Add("澳门");
            Cboprovince.Items.Add("台湾");
}

private void Cboprovince_SelectedIndexChanged(object sender, EventArgs e)
{
    CboCity.Items.Clear();
    try
    {
        switch (Cboprovince.Text.ToString())
        {
            case "河北":
                CboCity.Items.Add("石家庄");
                CboCity.Items.Add("唐山");
                CboCity.Items.Add("秦皇岛");
                CboCity.Items.Add("邯郸");
                CboCity.Items.Add("邢台");
                CboCity.Items.Add("保定");
                CboCity.Items.Add("张家口");
```

```
            CboCity.Items.Add("承德");
            CboCity.Items.Add("沧州");
            CboCity.Items.Add("廊坊");
            CboCity.Items.Add("衡水");
            break;
        case "山西":
            CboCity.Items.Add("太原");
            CboCity.Items.Add("大同");
            CboCity.Items.Add("阳泉");
            CboCity.Items.Add("长治");
            CboCity.Items.Add("晋城");
            CboCity.Items.Add("朔州");
            CboCity.Items.Add("晋中");
            CboCity.Items.Add("运城");
            CboCity.Items.Add("益州");
            CboCity.Items.Add("临汾");
            CboCity.Items.Add("吕梁");
            break;
        case "内蒙古":
            CboCity.Items.Add("呼和浩特");
            CboCity.Items.Add("包头");
            CboCity.Items.Add("乌海");
            CboCity.Items.Add("赤峰");
            CboCity.Items.Add("通辽");
            CboCity.Items.Add("鄂尔多斯");
            CboCity.Items.Add("呼伦贝尔");
            CboCity.Items.Add("巴彦佐尔");
            CboCity.Items.Add("乌兰察布");
            CboCity.Items.Add("兴安");
            CboCity.Items.Add("锡林郭勒");
            break;
        case "北京":
            ……//省略
            break;

    }
}
```

```
        catch (Exception ex)
        {
            MessageBox.Show(ex.Message.ToString());
        }
    }

    private void btnshow_Click(object sender, EventArgs e)
    {
        lblshow.Text = string.Format("你选择了{0}省{1}市", Cboprovince.Text, CboCity.Text);
    }
```

关键技术:

(1) 组合框是由一个文本框和一个列表框组成的,组合框又被称为弹出式菜单,用户使用时,单击文本框右侧的三角即可展开下拉列表;

(2) ComboBox 的相关属性、方法、事件的使用;

(3) 本示例利用 ComboBox 的 SelectedIndexChanged 事件,实现省市的联动。

【实验项目】

1.6 实现月日选择,当月份更改时,该月的天数也做相应修改,运行情况如图 1-7 所示。

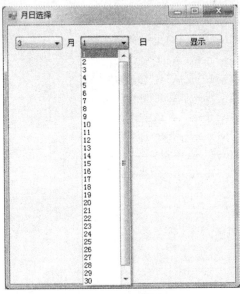

图 1-7 月日选择

1.1.6 树型组织机构示例程序

【实验1-6】 通过 TreeView 展现淮师各机构组成情况,通过 Button 实现对节点的折叠、展开、添加、删除等操作,程序运行效果如图1-8所示。

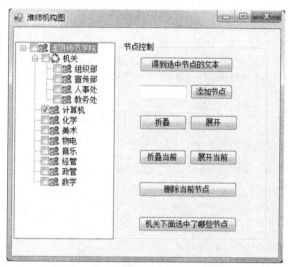

图1-8 淮师机构 TreeView 展示

相关控件:

各控件相关属性及事件如表1-6所示。

表1-6 控件属性设置及事件介绍

控件	名称	属性值	事件
Form	Form1	Text=淮师机构图	Load
Button	btnselecttxt	Text=得到选中节点的文本	Click
	btnaddnode	Text=添加节点	Click
	btncollapseAll	Text=折叠所有	Click
	btnexpandall	Text=展开所有	Click
	btncollapse	Text=折叠当前	Click
	btndelnode	Text=删除当前节点	Click
	btnexpand	Text=展开当前	Click
	btnselectednode	Text=选中了哪些机关节点	Click

(续表)

控件	名称	属性值	事件
TreeView	treeView 1	ImageList=imageList1	AfterCheck
		Nodes 属性设置节点	
GroupBox	groupBox1	Text=节点控制	
ImageList	imageList1	Images 属性作相关设置	

主要代码：

```
private void btnselecttxt_Click(object sender, EventArgs e)
{
    MessageBox.Show(treeView1.SelectedNode.Text);
}

private void btnaddnode_Click(object sender, EventArgs e)
{
    if (treeView1.SelectedNode != null)
    {
        treeView1.SelectedNode.Nodes.Add(textBox1.Text);
        textBox1.Text = "";
    }
    else
    {
        treeView1.Nodes.Add(textBox1.Text);
    }
}

private void Form1_Load(object sender, EventArgs e)
{
    treeView1.ExpandAll();
}

private void btncollapseAll_Click(object sender, EventArgs e)
{
    treeView1.CollapseAll();
```

```csharp
}

private void btnexpandall_Click(object sender, EventArgs e)
{
    treeView1.ExpandAll();
}

private void btncollapse_Click(object sender, EventArgs e)
{
    if (treeView1.SelectedNode != null)
    {
        treeView1.SelectedNode.Collapse();
    }
}

private void btnexpand_Click(object sender, EventArgs e)
{
    if (treeView1.SelectedNode != null)
    {
        treeView1.SelectedNode.Expand();
    }
}

private void btndelnode_Click(object sender, EventArgs e)
{
    treeView1.SelectedNode.Remove();
}

private void btnselectednode_Click(object sender, EventArgs e)
{
    string str = "选中的部门";
    foreach (TreeNode tn in treeView1.Nodes[0].Nodes[0].Nodes)
    {
        if (tn.Checked)
        {
            str = str + tn.Text;
        }
```

```
        }
        MessageBox.Show(str);
    }

    private void treeView1_AfterCheck(object sender, TreeViewEventArgs e)
    {
        foreach (TreeNode tn in e.Node.Nodes)
        {
            tn.Checked = e.Node.Checked;
        }
    }
```

关键技术：

(1) TreeView 控件用来显示信息的分级视图，如同 Windows 里的资源管理器的目录。TreeView 控件中的各项信息都有一个与之相关的 Node 对象。TreeView 显示 Node 对象的分层目录结构，每个 Node 对象均由一个 Label 对象和其相关的位图组成。在建立 TreeView 控件后，我们可以展开和折叠、显示或隐藏其中的节点。TreeView 控件一般用来显示文件和目录结构、文档中的类层次、索引中的层次和其他具有分层目录结构的信息。

(2) 本示例主要利用 Node 对象的 Checked 方法、TreeView 控件的 AfterCheck 事件实现 Node 的选择、判断。

1.2 图形图像程序设计基础

1.2.1 电子表示例程序

【实验 1-7】 设计一个与计算机时间和日期同步的电子表，程序运行效果如图如图 1-9 所示。

相关控件：

控件相关属性及事件如表 1-7 所示。

表 1-7 控件属性设置及事件介绍

控件	名称	属性值	事件
Form	Form1	Text=电子表	
Label	timeLabel	Text=""	
	weekLabel	Text=""	
Timer	watchTimer	Enabled=true InterVal=1000	Tick
PictureBox	backPictureBox	Image	

图 1-9 电子表

主要代码：

```
private void watchTimer_Tick(object sender, EventArgs e)
{
    DateTime dt = DateTime.Now;
    string weekday = "";

    switch (dt.DayOfWeek.ToString())
    {
```

```
            case "Monday":
                weekday = "星期一";
                break;
            case "Tuesday":
                weekday = "星期二";
                break;
            case "Wednesday":
                weekday = "星期三";
                break;
            case "Thursday":
                weekday = "星期四";
                break;
            case "Friday":
                weekday = "星期五";
                break;
            case "Saturday":
                weekday = "星期六";
                break;
            case "Sunday":
                weekday = "星期日";
                break;
        }
        weekLabel.Text = weekday;
        timeLabel.Text = dt.Hour.ToString("00") + ":" + dt.Minute.ToString("00") + ":" + dt.Second.ToString("00");
    }
```

关键技术：

(1) Timer 控件可根据应用程序需要按固定的时间间隔执行 Tick 事件中的代码，其应用面较广。

(2) 当前系统日期和时间可通过封装好的 DateTime 结构获得。

【实验项目】

1.7 在电子表程序中，实现点击表的右上按钮，则在表的时间区显示当前的年月日，并且 5 秒钟后自动恢复时间的显示的功能。

实验项目 1.7

1.2.2 线的绘制示例程序

【实验 1-8】 设计程序能动态绘制直线、曲线、椭圆、椭圆弧、填充椭圆等,程序运行效果如图如图 1-10 所示。

相关控件:

控件相关属性及事件如表 1-8 所示。

表 1-8 控件属性设置及事件介绍

控件	名称	属性值	事件
Form	Form1	Text=图形绘制	
GroupBox	groupBox1	Text=直线绘制区	
	groupBox2	Text=曲线绘制区	
	groupBox3	Text=圆形绘制区	
	groupBox4	Text=椭圆形绘制区	
	groupBox5	Text=圆弧绘制区	
	groupBox6	Text=填充椭圆绘制区	
PictureBox	pictureBoxline		MouseUp
			MouseDown
	pictureBoxcurve		MouseMove
	pictureBoxcircle		MouseUp
			MouseDown
	pictureBoxellipse		MouseUp
			MouseDown
	pictureBoxarc		MouseUp
			MouseDown
	pictureBoxfilledellipse		MouseUp
			MouseDown

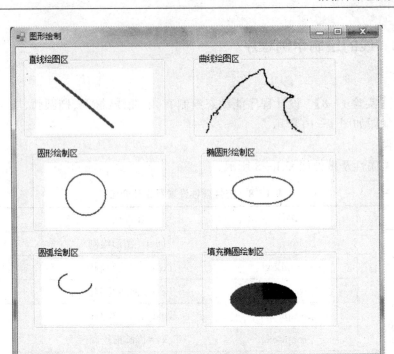

图 1-10 动态绘制图形

主要代码：

```
public partial class Form1 : Form
{
    Graphics g;
    int startX;
    int startY;
    public Form1()
    {
        InitializeComponent();
    }
    //绘制直线
    private void pictureBoxline_MouseUp(object sender, MouseEventArgs e)
    {
        g = pictureBoxline.CreateGraphics();
        Pen p = new Pen(Color.Red, 4);
        g.DrawLine(p, startX, startY, e.X, e.Y);
```

```csharp
}

private void pictureBoxline_MouseDown(object sender, MouseEventArgs e)
{
    pictureBoxline.BackColor = Color.Snow;
    startX = e.X;
    startY = e.Y;
}
//绘制曲线
private void pictureBoxcurve_MouseMove(object sender, MouseEventArgs e)
{
    g = pictureBoxcurve.CreateGraphics();
    Pen p = new Pen(Color.Blue, 2);
    g.DrawEllipse(p, e.X, e.Y, 1, 1);
}

private void pictureBoxcurve_MouseDown(object sender, MouseEventArgs e)
{
    pictureBoxcurve.BackColor = Color.Snow;
}
//绘制圆
private void pictureBoxcircle_MouseDown(object sender, MouseEventArgs e)
{
    pictureBoxcircle.BackColor = Color.Snow;
    startX = e.X;
    startY = e.Y;
}

private void pictureBoxcircle_MouseUp(object sender, MouseEventArgs e)
{
    Graphics g = pictureBoxcircle.CreateGraphics();
    g.Clear(Color.White);
    Pen p = new Pen(Color.Red, 2);
    g.DrawEllipse(p, startX, startY, Math.Abs(e.X - startX), Math.Abs(e.X - startX));
}
```

```csharp
//绘制椭圆
private void pictureBoxellipse_MouseDown(object sender, MouseEventArgs e)
{
    pictureBoxellipse.BackColor = Color.Snow;
    startX = e.X;
    startY = e.Y;
}

private void pictureBoxellipse_MouseUp(object sender, MouseEventArgs e)
{
    Graphics g = pictureBoxellipse.CreateGraphics();
    g.Clear(Color.White);
    Pen p = new Pen(Color.Red, 2);
    g.DrawEllipse(p, startX, startY, Math.Abs(e.X - startX), Math.Abs(e.Y - startY));
}

//绘制圆弧
private void pictureBoxarc_MouseDown(object sender, MouseEventArgs e)
{
    pictureBoxarc.BackColor = Color.Snow;
    startX = e.X;
    startY = e.Y;
}

private void pictureBoxarc_MouseUp(object sender, MouseEventArgs e)
{
    Graphics g = pictureBoxarc.CreateGraphics();
    g.Clear(Color.White);
    Pen p = new Pen(Color.Red, 2);
    g.DrawArc(p, startX, startY, 50, 30, 0, 230);
}

//绘制填充椭圆
private void pictureBoxfilledellipse_MouseDown(object sender, MouseEventArgs e)
{
    pictureBoxfilledellipse.BackColor = Color.Snow;
    startX = e.X;
```

第1章 Visual Studio 2010 C#程序设计基础

```
        startY = e.Y;
    }

    private void pictureBoxfilledellipse_MouseUp(object sender, MouseEventArgs e)
    {
        Graphics g = pictureBoxfilledellipse.CreateGraphics();
        g.Clear(Color.White);
        g.FillPie(new SolidBrush(Color.Red), startX, startY, 100, 50, 0, 270);
        g.FillPie(new SolidBrush(Color.Blue), startX, startY, 100, 50, 270, 90);
    }
}
```

关键技术：

(1) 作为图形设备接口的 GDI+ 使得应用程序开发人员在输出屏幕和打印机信息的时候无需考虑具体显示设备的细节，他们只需调用 GDI+ 库输出的类的一些方法即可完成图形操作，真正的绘图工作由这些方法交给特定的设备驱动程序来完成。GDI+ 使得图形硬件和应用程序相互隔离，使开发人员编写设备无关的应用程序变得非常容易。

(2) 本示例中用到了 GDI+ 各种绘图方法，如 FillPie、DrawArc、DrawEllipse、DrawLine 等，需掌握各种方法中各参数的含义。

1.2.3 验证码图像的生成示例程序

【实验 1-9】 设计如图 1-11 所示程序，动态生成验证码。

相关控件：

控件相关属性及事件如表 1-9 所示。

表 1-9 控件属性设置及事件介绍

控件	名称	属性值	事件
Form	Form1	Text=动态验证码	
GroupBox	groupBox1	Text=验证码	
PictureBox	pictureBox1		
Button	btnother		Click
	btnexit		Click

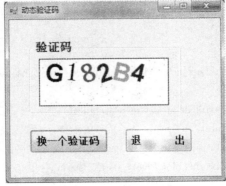

图 1-11 动态验证码

主要代码：

```
public partial class Form1 : Form
{
    private const int codelength = 6;
    private string randomcode = "";

    public Form1()
    {
        InitializeComponent();
    }
    //生成一定长度的验证码
    private string CreateRandomeCode(int length)
    {
        int rand;
        char code;
        string randomcode = string.Empty;
        System.Random random = new Random();
        for (int i = 0; i < length; i++)
        {
            rand = random.Next();
            if (rand % 3 == 0)
                code = (char)('A' + (char)(rand % 26));
            else
            {
                code = (char)('0' + (char)(rand % 10));
```

```csharp
        }
        randomcode += code.ToString();
    }
    return randomcode;
}
//创建随机码图片
private void createimage(string randomcode)
{
    if(randomcode==null || randomcode.Trim()==String.Empty)
    {
        return;
    }
    System.Drawing.Bitmap image = new Bitmap((int)Math.Ceiling(randomcode.Length * 35.0), 75);
    Graphics g = Graphics.FromImage(image);
    try
    {
        //绘制边框
        int randAngle = 30;
        g.Clear(Color.White);
        g.DrawRectangle(new Pen(Color.Black, 0), 0, 0, image.Width - 1, image.Height - 1);
        g.SmoothingMode = System.Drawing.Drawing2D.SmoothingMode.AntiAlias;
        Random rand = new Random();
        //背景噪点生成
        Pen blackPen = new Pen(Color.LightBlue, 0);
        for (int i = 0; i < 50; i++)
        {
            int x = rand.Next(0, pictureBox1.Width);
            int y = rand.Next(0, pictureBox1.Height);
            g.DrawRectangle(blackPen, x, y, 1, 1);
        }
        char[] chars = randomcode.ToCharArray();
        //拆散字符串成单个字符数组
        //定义文字居中
        StringFormat format = new StringFormat(StringFormatFlags.NoClip);
        format.Alignment = StringAlignment.Center;
        format.LineAlignment = StringAlignment.Center;
```

```
            //定义颜色
            Color[] c = { Color.Black, Color.Red, Color.LimeGreen, Color.MidnightBlue,
Color.Green, Color.Blue };
            //定义字体
            string[] font = { "Microsoft Sans Serif", "Arial", "宋体" };

            for (int i = 0; i < chars.Length; i++)
            {
                int cindex = rand.Next(6);
                int findex = rand.Next(3);
                Font f = new Font(font[findex], 30, System.Drawing.FontStyle.Bold);
                Brush b = new System.Drawing.SolidBrush(c[cindex]);
                Point dot = new Point(25, 25);
                float angle = rand.Next(-randAngle, randAngle);
                g.TranslateTransform(dot.X, dot.Y);
                g.RotateTransform(angle);
                g.DrawString(chars[i].ToString(), f, b, 1, 1, format);
                g.RotateTransform(-angle);
                g.TranslateTransform(2, -dot.Y);
            }
            this.pictureBox1.Width = image.Width;
            this.pictureBox1.Height = image.Height;
            pictureBox1.BackgroundImage = image;
        }
        catch (ArgumentException)
        {
            MessageBox.Show("创建图片错误");
        }
    }
    private void Form1_Load(object sender, EventArgs e)
    {
        randomcode = CreateRandomeCode(codelength);
        createimage(randomcode);
    }
    private void btnother_Click(object sender, EventArgs e)
    {
```

第 1 章　Visual Studio 2010 C#程序设计基础

```
            randomcode = CreateRandomeCode(codelength);
            createimage(randomcode);
        }
}
```

关键技术：
(1) 通过随机生成数的方法，随机生成一定长度的字符串。
(2) 利用 GDI+的绘图方法，实现随机字符串的绘制。

【实验项目】

1.8　结合实验示例 1-4、示例 1-9，设计如图 1-12 所示登录程序。

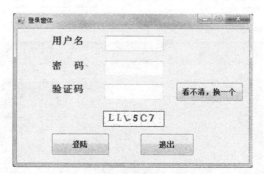

实验项目 1.8

图 1-12　带验证码登录程序

1.3　MP3 播放器示例程序

【实验 1-10】　设计如图 1-13 所示程序，利用 VS2010 提供控件实现 mp3 音乐的播放及相关播放控制。

相关控件：
控件相关属性及事件如表 1-10 所示。

表 1-10　表控件属性设置及事件介绍

控件	名称	属性值	事件
Form	Form1	Text=音乐播放器	
GroupBox	groupBox1	Text=播放控制	

（续表）

控件	名称	属性值	事件
ListBox	listBox1		MouseDoubleClick
Button	btnsetdir	Text=初始目录	Click
	btnexit	Text=退出	Click
	btnaddsong	Text=添加 Mp3 歌曲	Click
	btnplay	Text=暂停/播放	Click
	btndel	Text=删除	Click
	btnstop	Text=停止	Click
axWindows MediaPlayer	axWindowsMediaPlayer1		PlayStateChange

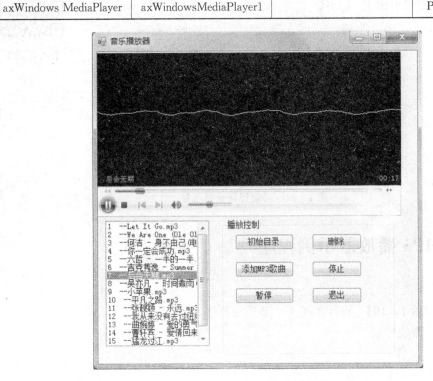

图 1-13　MP3 音乐播放器

主要代码：

```
public partial class Form1 : Form
{
    private string[] playlist = new string[10000];
```

```csharp
private int num;
private int state = 0;
private bool flag=false;
public Form1()
{
    InitializeComponent();
}


public void addfile(string path)      //自定义添加mp3歌曲方法
{
    if (num < 10000)
    {
        num++;
        playlist[num] = path;
    }
}

public void delfile(int selectnum)            //自定义删除歌曲方法
{
    for (int i = selectnum; i <= num - 1; i++)
    {
        playlist[i] = playlist[i + 1];
    }
    num--;
}

private void addfiles(string path, ListBox listBox1)
//自定义mp3歌曲目录方法
{
    DirectoryInfo dir = new DirectoryInfo(path);           //定义目录实例
    foreach (FileInfo f in dir.GetFiles("*.mp3"))
    {
        addfile(f.FullName);
        string strtmp = Convert.ToString(num);
        for (int i = 1; i <= 5 - strtmp.Length; i++)
        {
```

```csharp
                strtmp+=' ';
            }
            strtmp+="--"+f.Name;
            listBox1.Items.Add(strtmp);
        }
        foreach(DirectoryInfo f in dir.GetDirectories())
        {
            addfiles(f.FullName,listBox1);
        }
    }

    private void btnsetdir_Click(object sender, EventArgs e)        //初始目录
    {
        folderBrowserDialog1.SelectedPath = "d:\\";
        folderBrowserDialog1.ShowNewFolderButton = true;
        folderBrowserDialog1.Description = "请选择音乐文件目录";
        folderBrowserDialog1.ShowDialog();
        addfiles(folderBrowserDialog1.SelectedPath, listBox1);

        int selectone;
        if (listBox1.SelectedIndex < 0)
            selectone = 1;
        else
            selectone = listBox1.SelectedIndex + 1;
        if (listBox1.Items.Count < 0)
            listBox1.SelectedIndex = 0;
        axWindowsMediaPlayer1.URL = playlist[selectone];
    }

    private void Form1_Load(object sender, EventArgs e)
    {
        listBox1.Items.CopyTo(playlist, 0);
        num = 0;
    }

    private void btnaddsong_Click(object sender, EventArgs e)        //添加歌曲
```

```
            {
                    openFileDialog1.Filter = "*.mp3|*.mp3";
                    if (openFileDialog1.ShowDialog() == DialogResult.OK)
                    {
                        string path = openFileDialog1.FileName;
                        FileInfo f = new FileInfo(path);
                        addfile(f.FullName);
                        string strtmp = Convert.ToString(num);
                        for (int i = 1; i <= 5 - strtmp.Length; i++)
                            strtmp += ' ';
                        strtmp += "--" + f.Name;
                        listBox1.Items.Add(strtmp);
                    }
            }

            private void btnplay_Click(object sender, EventArgs e)    //播放或暂停
            {
                int selectone;
                if (state == 0 || state == 1)
                {
                    if (listBox1.SelectedIndex < 0)
                        selectone = 1;
                    else
                        selectone = listBox1.SelectedIndex + 1;
                    if (listBox1.Items.Count < 0)
                        listBox1.SelectedIndex = 0;
                    axWindowsMediaPlayer1.URL = playlist[selectone];
                }
                else if (state == 3 && btnplay.Text == "暂停")
                    axWindowsMediaPlayer1.Ctlcontrols.pause();
                else if(state==2 && btnplay.Text=="播放")
                    axWindowsMediaPlayer1.Ctlcontrols.play();
            }

            private void btndcl_Click(object sender, EventArgs e)    //删除歌曲
            {
                if (listBox1.SelectedIndex >= 0)
```

```
            {
                    delfile(listBox1.SelectedIndex + 1);
                    listBox1.Items.RemoveAt(listBox1.SelectedIndex);
            }
    }

    private void btnstop_Click(object sender, EventArgs e)        //停止播放
    {
            axWindowsMediaPlayer1.Ctlcontrols.stop();
    }
    private void btnexit_Click(object sender, EventArgs e)        //退出
    {
        Application.Exit();
    }
    private void listBox1_MouseDoubleClick(object sender, MouseEventArgs e)
    {
        int selectone;
        if (listBox1.SelectedIndex < 0)
            selectone = 1;
        else
            selectone = listBox1.SelectedIndex + 1;
        if (listBox1.Items.Count < 0)
            listBox1.SelectedIndex = 0;
        axWindowsMediaPlayer1.URL = playlist[selectone];
    }
    private void axWindowsMediaPlayer1_PlayStateChange(object sender, AxWMPLib._WMPOCXEvents_PlayStateChangeEvent e)
    {
        if (e.newState == 0)
            btnplay.Text = "播放";
        else if (e.newState == 1)
            btnplay.Text = "播放";
        else if (e.newState == 2)
            btnplay.Text = "播放";
        else if (e.newState == 3)
            btnplay.Text = "暂停";
        state = e.newState;
    }
```

第 1 章　Visual Studio 2010 C# 程序设计基础　　　　　　　　　　　　　　　35

关键技术：

（1）axWindowsMediaPlayer 控件的使用。

（2）listBox 控件显示载入的音乐文件，通过 listBox1_MouseDoubleClick 事件直接切换音乐。

（3）button 控件控制播放、暂停、停止、删除、退出等等。

【实验项目】

1.9　使用 Shockwave Flash Object 实现 Flash 动画的控制，如：播放 Flash、停止 Flash、浏览 Flash、第一帧、下一帧、上一帧等。

实验项目 1.9

1.4　文件操作示例程序

【实验 1-11】　通过文件打开对话框打开文本文件。统计文本文件中指定字符串的个数，并以大字红色的形式在文本中显示指定字符串。

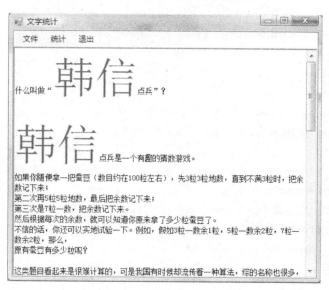

图 1-14　文字统计

相关控件：

用控件及相关属性如表 1-11 所示。

表 1-11　控件属性设置及事件介绍

控件	名称	属性值	事件
Form	Form1	Text=文字统计	
RichTextBox	RichTextBox1		
MenuStrip	菜单	文件	Click
	菜单	统计	Click
	菜单	退出	Click
OpenFileDialog	OpenFileDialog1		

主要代码：

```csharp
private void 文件ToolStripMenuItem_Click(object sender, EventArgs e)
{
    if (openFileDialog1.ShowDialog() == DialogResult.OK)
        richTextBox1.LoadFile(openFileDialog1.FileName, RichTextBoxStreamType.PlainText);
}

private void 统计ToolStripMenuItem_Click(object sender, EventArgs e)
{
    string findStr;
    int loc, sum = 0, length;
    StringInputDialog inputDialog = new StringInputDialog();
    if (inputDialog.ShowDialog() == DialogResult.OK)
    {
        findStr = inputDialog.StatString;
        length = findStr.Length;
        richTextBox1.Select(0, richTextBox1.Text.Length);
        richTextBox1.SelectionFont = new System.Drawing.Font("宋体", 9, FontStyle.Regular);
        richTextBox1.SelectionColor = Color.Black;
        loc = richTextBox1.Find(findStr);
        while (loc! = -1)
        {
            sum
            richTextBox1.Select(loc, length);
            richTextBox1.SelectionFont = new Font("宋体", 50, FontStyle.Regular);++;
```

```
                richTextBox1.SelectionColor = Color.Red;
                loc = richTextBox1.Find(findStr, loc + length, RichTextBoxFinds.MatchCase);
            }
            MessageBox.Show((sum!=0?("文中含字符串""+findStr+""的位置共计"+sum.ToString()+"个"):("查无""+findStr+""串!")),"统计结果",MessageBoxButtons.OK,MessageBoxIcon.Information,MessageBoxDefaultButton.Button1);
        }
    }

    private void 退出ToolStripMenuItem_Click(object sender, EventArgs e)
    {
        this.Close();
    }

    private void Form1_FormClosing(object sender, FormClosingEventArgs e)
    {
        if (MessageBox.Show("真要关闭应用程序吗?","提示",MessageBoxButtons.YesNo,MessageBoxIcon.Question,MessageBoxDefaultButton.Button1) == DialogResult.No)
            e.Cancel = true;
    }
```

关键技术：

RichTextBox 控件是一个功能非常强大的文本处理控件，本例中给出了对选中字符串进行字体大小和颜色修改的方法，其他功能读者可查阅帮助文件。

【实验项目】

1.10 模仿设计 Windows 操作系统中提供的记事本软件。

1.11 18 位身份证号码正误验证。18 位身份证标准在国家质量技术监督局于 1999 年 7 月 1 日实施的 GB11643-1999《公民身份号码》中做了明确规定。公民身份号码是特征组合码，由 17 位数字本体码和 1 位校验码组成。排列顺序从左至右依次为：6 位数字地址码，8 位数字出生日期码，3 位数字顺序码和 1 位校验码。其含义如下：

实验项目 1.10 和 1.11

地址码：表示编码对象常住户口所在县(市、旗、区)的行政区划代码，按 GB/T2260 的规定执行。

出生日期码：表示编码对象出生的年、月、日，按 GB/T7408 的规定执行，年、月、日分别用 4 位、2 位、2 位数字表示，之间不用分隔符。

顺序码：表示在同一地址码所标识的区域范围内，对同年、同月、同日出生的人编定的顺序号，顺序码的奇数分配给男性，偶数分配给女性。

校验的计算方式：对前 17 位数字本体码加权求和，公式为：$S = \sum_{i=1}^{17} A_i \times W_i$，其中 A_i 表示第 i 位置上的身份证号码数字值（从左到右），W_i 表示第 i 位置上的加权因子，其各位对应的值依次为：7 9 10 5 8 4 2 1 6 3 7 9 10 5 8 4 2。再以 11 对计算结果 S 取模。根据模的值得到对应的校验码，对应关系为：

    Y 值：0 1 2 3 4 5 6 7 8 9 10

    校验码：1 0 X 9 8 7 6 5 4 3 2

1.5 正则表达式示例程序

【实验 1-12】 通过文本框输入电话号码，点击验证 Button，可以通过正则表达式判断该电话号码是否可以通过验证。示例效果如图 1-12 所示。

相关控件：

所用控件及相关属性如表 1-12 所示。

表 1-12 控件属性设置及事件介绍

控件	名称	属性值	事件
Form	Form1	Text=正则表达式验证	
TextBox	txttelno	Text=""	
Label	label1	Text="请输入手机号码"	
	label2	Text=""	
Button	btncheck	Text="号码验证"	Click
	btnsetnull	Text="重输号码"	Click

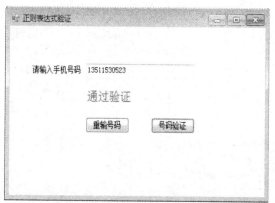 

图 1-12 号码验证

主要代码：

```
public static bool IsValidMobileNo(string MobileNo)
{
    //电信手机号码正则
    string dianxin = @"^133\d{8}|153\d{8}|180\d{8}|189\d{8}$";
    if (Regex.IsMatch(MobileNo, dianxin))
    {
        return true;
    }
    else
    {
        //联通手机号正则
        string liantong = @"^1[358][01256]\d{8}$";
        if (Regex.IsMatch(MobileNo, liantong))
        {
            return true;
        }
        else
        {
            //移动手机号正则
            string yidong = @"^(13[456789]\d{8}|15[012789]\d{8}|18[278]\d{8}|147\d{8})$";
            if (Regex.IsMatch(MobileNo, yidong))
            {
```

```
                return true;
        }
        else
                return false;
    }
}
private void btncheck_Click(object sender, EventArgs e)
{
    if(IsValidMobileNo(txttelno.Text.Trim()))
    {
        label2.Text="通过验证";
    }
    else
    {
        label2.Text="号码有误";
    }
}
private void btnsetnull_Click(object sender, EventArgs e)
{
    txttelno.Text = "";
}
```

关键技术：

（1）移动号段：134、135、136、137、138、139、147、150、151、152、157、158、159、182、187、188；联通号段：130、131、132、155、156、185、186；电信：133、153、180、189。

（2）掌握正则表达式的表示、判断方法。

【实验项目】

1.12 使用正则表达式实现电子邮箱的验证。

实验项目1.12

第 2 章 SQL Server 2008 数据库应用基础

SQL Server 2008 数据库管理系统是微软公司推出的新一代具有里程碑性质的企业级数据库产品,在安全性、高可靠性、性能、扩展性、可管理性等方面较先前的版本均有极大的提高,成为微软下一代的数据管理与商业智能平台。

本章从基本的数据库创建开始,结合王珊和萨师煊编著的《数据库系统概论》(第四版)教材(以下简称理论教材)中的内容,通过上机实验学习并实践 SQL 语句、视图、索引、数据完整性、存储过程、触发器、安全控制等数据库的基础知识和 SQL Server 2008 数据库管理系统的使用。

2.1 数据库创建与管理

2.1.1 数据库的创建与删除

SQL Server 2008 数据库管理软件提供了非常友好的用户界面,使用方便,大多数数据库管理均可在 SQL Server Management Studio 中完成。

【实验 2-1】 在 SQL Server 2008 中创建 exampleDB 数据库。

实验步骤:
方法 1:使用 Management Studio 管理工具
(1) 打开 Management Studio 软件,在对象资源管理器中对"数据库"点击鼠标右键。
(2) 选择"新建数据库",在对话框中输入数据库名称 exampleDB。
(3) 对新创建的 exampleDB 数据库点击鼠标右键,完成"重命名"或"删除"操作。
(4) 对 exampleDB 数据库点击鼠标右键,选择"属性",查看数据库文件 exampleDB. mdf 和 example_log. ldf 所存储的位置,到文件夹下找到两个文件,其中保存了数据库中的数据。

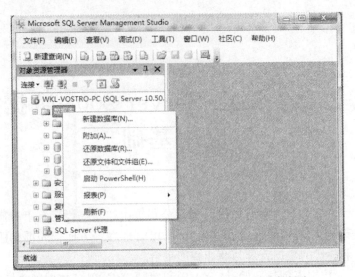

图 2-1　SQL Server Management Studio 创建数据库

方法 2：使用 SQL 语句

（1）在 Management Studio 中，点击工具栏的"新建查询"按钮，产生 SQLQuery1.sql 查询文件编辑窗口。

（2）在编辑窗口输入 CREATE DATABASE exampleDB 语句，点击"！"按钮，执行创建数据库语句。选中对象资源管理器窗口，按"F5"或对"数据库"项点击鼠标右键选择"刷新"，出现新建的数据库 exampleDB。如图 2-2 所示。

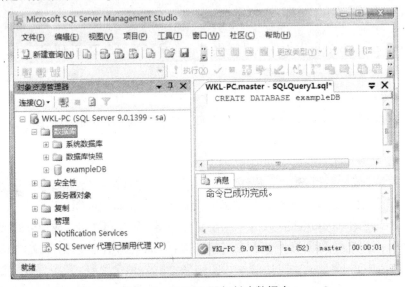

图 2-2　用 SQL 语句创建数据库

(3) 查阅数据库状态信息。在编辑窗口执行系统存储过程 EXEC sp_helpdb exampleDB。

(4) 数据库改名。在查询编辑窗口输入 ALTER DATABASE exampleDB MODIFY NAME=myDB 语句，选中该语句点击执行按钮，exampleDB 数据库被改为 myDB。

另一种为数据库改名的方法是执行系统存储过程，与上面语句效果等价的命令是 EXEC sp_renamedb 'exampleDB','myDB'。

(5) 删除数据库。执行语句 DROP DATABASE myDB。

(6) 保存查询文件。选中查询编辑窗口，从"文件"选择"保存 SQLQuery1.sql"功能或对编辑窗口的标题点击右键选"保存 SQLQuery1.sql"。

【实验项目】

2.1 用菜单方式和 SQL 语句方式分别创建、删除供应商－零件－工程项目数据库（命名为 SPJ），参见理论教材第二章练习 5。

2.1.2 认识 SQL Server 数据库

SQL Server 2008 为管理的需要设计了 4 个系统数据库，分别是 master,model,msdb 和 tempdb,如图 2-3 所示。这些数据库是系统的重要组成部分，由系统自动维护，用户不能修改其中的信息。每个数据库的主要作用如下：

master 数据库记录了 SQL Server 系统的所有系统级信息。包括实例范围的元数据（例如登录帐户）、端点、链接服务器和系统配置设置。master 数据库记录了所有其他数据库是否存在以及这些数据库文件的位置。另外，master 还记录了 SQL Server 的初始化信息。因此，master 数据库若不能正常运行，则 SQL Server 将无法启动。

model 数据库用作在 SQL Server 实例上创建的所有数据库的模板。当发出 CREATE DATABASE SQL 语句时，将通过复制 model 数据库中的内容来创建数据库的第一部分，再用空页填充新数据库的剩余部分。

msdb 数据库由 SQL Server 代理用来计划警报和作业。

tempdb 数据库是连接到 SQL Server 实例的所有用户都可用的全局资源，它保存所有临时表和临时存储过程。另外，它还用来满足所有其他临时存储要求，例如：存储 SQL Server 生成的工作表。每次启动 SQL Server 时，都要重新创建 tempdb，以便系统启动时，该数据库总是空的。在断开联接时会自动删除临时表和存储过程，并且在系统关闭后没有活动连接。因此 tempdb 中不会有什么内容从一个 SQL Server 会话保存到另一个会话。

在系统数据库之外，SQL Server 2008 还建立了一个系统使用的只读数据库，称为 Resource 数据库。它包含了 SQL Server 2008 中的所有系统对象，其主要作用是系统可比较轻松快捷地升级到新的 SQL Server 版本。

用户创建的数据库中含有表、数据库关系图、视图、可编程性、安全性等内容。

表是包含数据库中的所有数据的对象,是数据库的基础。

数据库关系图以图形方式显示数据库的结构。使用数据库关系图可以创建和修改表、列、关系和键。此外,还可以修改索引和约束。

可编程性中含有存储过程、函数、触发器等与数据库应用开发相关的项目。

图 2-3 数据库的组成

【实验项目】

2.2 从你的系统中找到用于 exampleDB 数据库的二个文件,复制到移动存储设备。注:复制前先对相应数据库名称右击,从弹出菜单中选择"任务"/"脱机"。

2.2 表的操作与视图

2.2.1 创建表

【实验 2-2】 在 exampleDB 数据库中,创建 Student,Course,SC 数据表。

实验步骤：

方法 1：使用工具

（1）在 exampleDB 数据库下，右击"表"，选择"新建表"。

（2）在列名、数据类型栏分别输入下列字段和数据类型。点击 Sno 行左端按钮，选择"设置主键"。如图 2-4 所示。

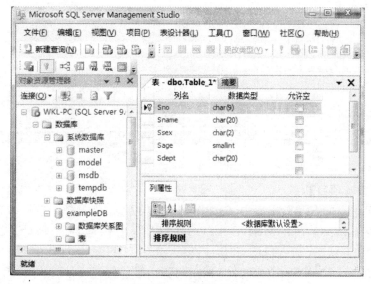

图 2-4 创建数据表

字段	数据类型
Sno	char(9),
Sname	char(20),
Ssex	char(2),
Sage	smallint,
Sdept	char(20),

（3）以 Student 名称保存表。

方法 2：使用 SQL 语句

（1）右击 exampleDB 数据库，创建新查询，输入下列语句并执行。分别建立选课表（Course）和成绩表（SC）

```
CREATE TABLE Course
 (Cno Char(4) PRIMARY KEY,
 Cname CHAR(40),
 Cpno CHAR(4) REFERENCES Course(Cno),
```

```
Ccredit SMALLINT);

CREATE TABLE SC
(Sno CHAR(9),
Cno CHAR(4) PRIMARY KEY(Sno,Cno),
Grade smallint,
FOREIGN KEY (Sno) REFERENCES Student(Sno),
FOREIGN KEY (Cno) REFERENCES Course(Cno),
);
```

（2）设置 Course 中 Cpno 是外码，被参照列是 Cno。使用表修改语句。

```
ALTER TABLE Course
WITH NOCHECK ADD FOREIGN KEY (Cpno) REFERENCES Course(Cno)
```

【实验项目】

2.3 在已建立的 SPJ 数据库中，创建下列四个关系表，参见理论教材第二章练习 5。

供应商表 S(SNO, SNAME, STATUS, CITY)，由供应商代码（SNO）、供应商名（SNAME）、供应商状态（STATUS）和供应商所在城市（CITY）4 个字段组成。

零件表 P(PNO, PNAME, COLOR, WEIGHT)，由零件代码（PNO）、零件名（PNAME）、颜色（COLOR）和重量（WEIGHT）4 个字段组成。

工程项目表 J(JNO, JNAME, CITY)，由工程项目代码（JNO），工程项目名（JNAME），工程项目所在城市（CITY）组成。

供应情况表 SPJ(SNO, PNO, JNO, QTY)，由供应商代码（SNO），零件代码（PNO），工程项目代码（JNO），供应数量（QTY）组成。

每个字段所选的数据类型和长度请读者根据实际给予合理设定。

2.2.2　索引

【实验 2-3】 为 exampleDB 数据库中的表建立索引。

实验步骤：
方法 1：使用工具
（1）右击 Student 表，选择"修改"，打开表设计器。
（2）在表设计器左端行按钮上右击，选择"索引"/"键"，出现对话框。在对话框中完成添加、修改和删除索引等操作。

方法2:使用SQL语句

(1) 在查询窗口,输入下列语句并执行,为三个表分别添加一个索引。

```
CREATE UNIQUE INDEX Stusno ON Student(Sno);
CREATE UNIQUE INDEX Coucno ON Course(Cno);
CREATE UNIQUE INDEX SCno ON SC(Sno ASC,Cno DESC);
```

(2) 在查询窗口,执行下列语句将删除上面建立的三个索引。

```
DROP INDEX Student.Stusno;
DROP INDEX Course.Coucno;
DROP INDEX SC.SCno;
```

【实验项目】

2.4 使用工具为SPJ数据库中的S表建立按城市升序的索引。用SQL语句为SPJ创建供应商代码升序、零件代码升序和工程项目代码升序的索引。删除建立的索引。

2.2.3 输入数据

【实验2-4】 向exampleDB数据库各个表中添加数据。

实验步骤:

方法1:使用编辑工具

展开exampleDB数据库节点,右击Student表,选择"编辑前200行",在编辑窗口中直接输入数据。

方法2:使用SQL语句

通过查询窗口,输入下列语句并执行。注:如果出现错误,查看Course表是否有Cpno外键。若有删除之,再输入Course表和SC表的数据。

```
INSERT INTO Student
VALUES ('200215121','李勇','男',20,'CS');
INSERT INTO Student(Sno,Sname,Ssex,Sage,Sdept)
VALUES ('200215122','刘晨','女',19,'IS');
INSERT INTO Student(Sno,Sname,Ssex,Sage,Sdept)
VALUES ('200215123','王敏','女',18,'MA');
INSERT INTO Student(Sno,Sname,Ssex,Sage,Sdept)
VALUES ('200215125','张立','男',19,'IS');
```

```
INSERT INTO Course
VALUES ('1','数据库','5',4);
INSERT INTO Course
VALUES ('2','数学','',2);
INSERT INTO Course
VALUES ('3','信息系统','1',4);
INSERT INTO Course
VALUES ('4','操作系统','6',3);
INSERT INTO Course
VALUES ('5','数据结构','7',4);
INSERT INTO Course
VALUES ('6','数据处理','',2);
INSERT INTO Course
VALUES ('7','PASCAL语言','6',4);
INSERT INTO SC
VALUES ('200215121','1',92);
INSERT INTO SC
VALUES ('200215121','2',85);
INSERT INTO SC
VALUES ('200215121','3',88);
INSERT INTO SC
VALUES ('200215122','2',90);
INSERT INTO SC
VALUES ('200215122','3',80);
```

【实验项目】

2.5 打开 SPJ 数据库,向表中分别添加下列数据。

S 表

SNO	SNAME	STATUS	CITY
S1	精益	20	天津
S2	盛锡	10	北京
S3	东方红	30	北京
S4	丰泰盛	20	天津
S5	为民	30	上海

P 表

PNO	PNAME	COLOR	WEIGHT
P1	螺母	红	12
P2	螺栓	绿	17
P3	螺丝刀	蓝	14
P4	螺丝刀	红	14
P5	凸轮	蓝	40
P6	齿轮	红	30

J 表

JNO	JNAME	CITY
J1	三建	北京
J2	一汽	长春
J3	弹簧厂	天津
J4	造船厂	天津
J5	机车厂	唐山
J6	无线电厂	常州
J7	半导体厂	南京

SPJ 表

SNO	PNO	JNO	QTY
S1	P1	J1	200
S1	P1	J3	100
S1	P1	J4	700
S1	P2	J2	100
S2	P3	J1	400
S2	P3	J2	200
S2	P3	J4	500
S2	P3	J5	400
S2	P5	J1	400
S2	P5	J2	100

(续表)

SNO	PNO	JNO	QTY
S3	P1	J1	200
S3	P3	J1	200
S4	P5	J1	100
S4	P6	J3	300
S4	P6	J4	200
S5	P2	J4	100
S5	P3	J1	200
S5	P6	J2	200
S5	P6	J4	500

2.2.4 数据查询

【实验2-5】 在exampleDB数据库上建立新查询，输入并执行查询语句，观察结果。查询语句选自《数据库系统概论》理论教材第3章例题，含义可参考教材。

实验步骤：

右击exampleDB数据库，创建新查询。输入下列各语句并执行。

```
SELECT *
FROM Student

SELECT Sname,2004-Sage
FROM STUDENT

SELECT Sname,'Year of Birth:',2004-Sage,LOWER(Sdept)
FROM student;

SELECT DISTINCT Sno
FROM SC;

SELECT distinct sno
FROM SC
```

WHERE grade<100;

SELECT sname,sdept,sage
FROM student
WHERE sage BETWEEN 20 AND 23;

SELECT sname,sdept,sage
FROM student
WHERE sage NOT BETWEEN 20 AND 23;

SELECT Sname,Ssex
FROM student
WHERE Sdept IN ('IS','MA','CS')

SELECT Sname,Sno,Ssex
FROM student
WHERE Sname like '刘%'

SELECT Sname,Sno
FROM student
WHERE Sname like '_阳%'

SELECT Cno,Ccredit
FROM Course
WHERE Cname like 'DB_Design' ESCAPE '\'

SELECT Sno,Cno
FROM SC
WHERE Grade IS NULL;

SELECT sno,Grade
FROM sc
WHERE cno='3'
ORDER BY grade DESC

```sql
SELECT COUNT(*)
FROM student;

SELECT COUNT(DISTINCT Sno)
FROM SC;

SELECT AVG(grade)
FROM SC
WHERE Cno='1';

SELECT MAX(grade)
FROM SC
WHERE Cno='1'

SELECT SUM(Ccredit)
FROM SC,Course
WHERE Sno='200215121' AND SC.Cno=Course.Cno

SELECT Cno,COUNT(Sno)
FROM SC
GROUP BY Cno

SELECT Sno
FROM SC
GROUP BY Sno
HAVING COUNT(*)>2;

SELECT F.Cno,S.Cpno
FROM Course F,Course S
WHERE F.Cpno=S.Cno;

SELECT Student.sno,Sname,Ssex,Sage,Sdept,Cno,Grade
FROM Student,SC
WHERE Student.Sno *= SC.Sno
```

SELECT Student.sno,Sname,Ssex,Sage,Sdept,Cno,Grade
FROM Student LEFT OUTER JOIN SC ON Student.Sno=SC.Sno

SELECT Student.sno,Sname,Ssex,Sage,Sdept,Cno,Grade
FROM Student LEFT JOIN SC ON Student.Sno=SC.Sno

SELECT Student.Sno, Sname,Cname,Grade
FROM Student,SC,Course
WHERE Student.sno=SC.sno AND SC.Cno=Course.Cno;

SELECT Sno,Sname
FROM Student
WHERE Sno IN
    (SELECT Sno
    FROM SC
    WHERE Cno IN
        (SELECT Cno
        FROM Course
        WHERE Cname='信息系统'));
SELECT Sname,Sage
FROM Student
WHERE Sage < ALL
    (SELECT Sage
    FROM Student
    WHERE Sdept ='CS') AND
    Sdept<>'CS'

SELECT Sname
FROM Student
WHERE EXISTS
    (SELECT *
    FROM SC
    WHERE Sno = Student.Sno AND Cno='1');

SELECT Sname

```
FROM Student
WHERE NOT EXISTS
    (SELECT *
     FROM Course
     WHERE NOT EXISTS
        (SELECT *
         FROM SC
         WHERE Sno = Student.Sno AND Cno=Course.Cno));

SELECT DISTINCT Sno
FROM SC SCX
WHERE NOT EXISTS
    (SELECT *
     FROM SC SCY
     WHERE SCY.Sno = '200215122' AND
        NOT EXISTS
        (SELECT *
         FROM SC SCZ
         WHERE SCZ.Sno=SCX.Sno AND SCZ.Cno=SCY.Cno));

SELECT *
FROM Student
WHERE Sdept='CS'
UNION
SELECT *
FROM Student
WHERE Sage<=19

SELECT *
FROM Student
WHERE Sdept='CS'
INTERSECT
SELECT *
FROM Student
WHERE Sage<=19
```

```
SELECT Sno
FROM SC
WHERE Cno='1' AND Sno IN
    (SELECT Sno
     FROM SC
     WHERE Cno='2');

SELECT *
FROM Student
WHERE Sdept='CS' AND Sage>19;
```

【实验项目】

2.6 使用 SQL 语句完成以下各项操作。参见理论教材第 3 章练习。

1. 求供应工程 J1 零件的供应商代码。
2. 求供应工程 J1 零件 P1 的供应商代码。
3. 求没有使用天津供应商生产的红色零件的工程项目代码。
4. 求至少用了供应商 S1 所供应的全部零件的工程项目代码。
5. 找出所有供应商的名称和所在城市。
6. 找出所有零件的名称、颜色、重量。
7. 找出使用供应商 S1 所供应零件的工程项目代码。
8. 找出工程项目 J2 使用的各种零件的名称及其数量。
9. 找出上海厂商供应的所有零件代码。
10. 找出使用上海产的零件的工程项目名称。
11. 找出没有使用天津产的零件的工程代码。
12. 把全部红色零件的颜色改成蓝色。
13. 由 S5 供给 J4 的零件 P6 改为由 S3 供应，请作必要的修改。
14. 从供应商关系中删除 S2 的记录，并从供应情况关系中删除相应的记录。
15. 请将(S2,P4,J6,200)插入供应情况关系。

实验项目 2.6

2.2.5 关系图

关系图是以直观的方式表示数据源视图所包含的表、关系、命名查询和命名计算。为使数据库设计可视化，可创建一个或多个关系图，以直观的方式呈现数据库中的部分或全部表、列、键和关系。

数据库关系图设计器采用直观的图形化的设计方式,允许创建、编辑或删除表、列、键、索引、关系和约束等各种操作,功能较强大。

【实验 2-6】 利用数据库关系图设计器,在 exampleDB 数据库中创建数据库关系图。

实验步骤:

(1) 在对象资源管理器中,展开 exampleDB 数据库节点,右击"数据库关系图",选择"新建数据库关系图"。

(2) 在对话框中,点击"添加"按钮,将 Student,SC,Course 三个数据表添加到关系图中。如图 2-5 所示。

(3) 右击关系图中的对象,在弹出菜单中选择各种功能,熟悉工具的使用。

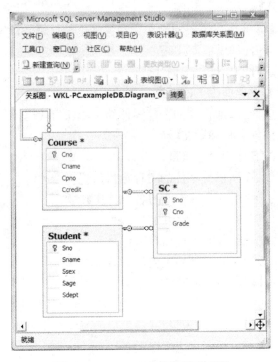

图 2-5 exampleDB 数据库关系图

【实验项目】

2.7 为 SPJ 数据库的四个表添加关系图。如果表没有设置主键,在关系图中为表设置主键。为 SPJ 表的前三个字段设置外键。

2.2.6 视图

视图是一个虚拟表,其内容是从相关表导出。同表一样,视图包含一系列带有名称的列和行数据。行和列数据来自由定义视图的查询所引用的表,并且在引用视图时动态生成。通过视图进行查询没有任何限制,进行数据修改时限制也较少。

在 SQL Server 2008 中,视图分为标准视图、索引视图和分区视图。标准视图组合了一个或多个表中的数据,是一般意义上的视图;索引视图是被具体化了的视图,即它已经过计算并存储,它可以显著提高某些类型查询的性能,但不太适于经常更新的基本数据集;分区视图在一台或多台服务器间水平连接一组成员表中的分区数据。

【实验 2-7】 在 exampleDB 数据库中创建视图。

实验步骤:

方法 1:使用工具

(1)展开 exampleDB 数据库节点,右击"视图",选择"新建视图"。

(2)添加 Student 表并关闭对话框,选择 Sno,Sname,Sage 字段,在 SELECT 语句最后输入 WHERE Sdept='IS'语句,点击"!"按钮执行 SQL,窗口下端显示执行结果。

(3)点击关闭,以 IS_Student 名称保存建立的视图。展开视图节点,右击保存的视图,选择"删除",删除已建立的 IS_Student 视图。

方法 2:使用 SQL 语句

选中 exampleDB 数据库节点,新建查询,在查询编辑窗口中编辑并执行下面语句。

```
CREATE VIEW IS_Student
AS
    SELECT Sno,Sname,Sage
    FROM Student
    WHERE Sdept='IS'
    WITH CHECK OPTION


SELECT * FROM IS_Student


UPDATE IS_Student
SET Sname='李勇'
WHERE Sno='95003'


INSERT INTO IS_Student
```

```
VALUES('95033','赵新',20)

DROP VIEW IS_Student

CREATE VIEW S_G(Sno,Gavg)
AS
SELECT Sno,AVG(Grade)
FROM SC
GROUP BY Sno

SELECT  *  FROM S_G
```

【实验项目】

2.8 在 SPJ 数据库中,为三建工程项目建立一个供应情况视图,包括供应商代码、零件代码、供应数量。针对该视图完成下列查询:

(1) 找出三建工程项目使用的各种零件代码及其数量;

(2) 找出供应商 S1 的供应情况。

2.3 数据库安全性

2.3.1 服务器身份验证模式

登录 SQL Server 2008 数据库需要通过身份验证,认证分为 Windows 身份验证和混合身份验证二种模式。如图 2-6 所示。

使用 Windows 身份验证模式时,用户通过 Microsoft Windows 用户帐户连接,SQL Server 使用 Windows 操作系统中的信息验证帐户名和密码。这是默认的身份验证模式,比混合模式安全。

混合身份验证模式是指 SQL Server 和 Windows 身份验证模式。提供 SQL Server 身份验证是为了兼容性,因为 7.0 及先前的版本只有这种验证模式。

使用"混合模式身份验证"时,要求输入系统管理员(sa)登录名和密码。密码是抵御入侵者的第一道防线,设置强密码对于系统安全是绝对必要的,不要设置空的或弱的 sa 密码。SQL Server 密码可包含 1 到 128 个字符,包括字母、符号和数字的任意组合。

图 2-6 服务器的安全性设置

【实验 2-8】修改服务器身份验证模式。

实验步骤：

（1）在对象资源管理器中，右击数据库服务器，选择"属性"。

（2）选择"安全性"页，在服务器身份验证栏下，选择新的服务器身份验证模式，再单击"确定"。修改在重新启动 SQL Server 2008 后方能生效。

2.3.2 登录名与服务器角色

SQL Server 2008 中使用角色管理对象和数据安全性，角色是一组访问权限的集合，通过简单地把用户分配到某个角色中，就能将这一组访问权限一起指派给用户。角色分为两类：服务器角色和数据库角色。

服务器角色是一个固定角色，用于提供对 Analysis Services 实例的管理员访问权限。如创建登录账户和创建链接服务器。服务器角色无法创建，只能使用系统定义好的 8 个服务器角色。每个角色的权限见表2-1。

表2-1 服务器角色的权限

角色名称	权限
sysadmin	可以在服务器中执行任何活动。
Serveradmin	可以设置服务器范围的配置选项,关闭服务器。
Setupadmin	可以管理链接服务器和启动过程。
Securityadmin	可以管理登录和CREATE DATABASE权限,还可以读取错误日志和更改密码。
Processadmin	可以结束在服务器执行个体中运行的进程。
Dbcreator	可以创建、更改、删除以及还原任何数据库。
Diskadmin	可以管理磁盘文件。
Bulkadmin	可以执行BULK INSERT语句,执行批量日志处理。

登录名是用户登录服务器的账号,用户必须用登录名进行连接,以获取 SQL Server 服务器的访问权限。登录名属于服务器层面,并不能用来访问服务器中的数据库,必须拥有数据库的用户账户才能访问数据库中数据。

【实验2-9】在数据库系统中创建新的登录名。

实验步骤:

·方法1:使用工具

(1)在"对象资源管理器"中展开服务器,再展开"安全性",右击"登录名",选择"新建登录名"。弹出新建登录名对话框,如图2-7所示。

(2)若选择"Windows 身份验证",登录名只能通过点击搜索按钮进行选择。若选择"SQL Server 身份验证",登录名直接输入,并输入密码。选择"SQL Server 身份验证",输入登录名 LiLi,并输入密码 123456。

(3)在对话框中选择"用户映射",通过复选框选择此登录名映射的数据库 exampleDB,系统自动在被选择的数据库中建立用户账户 LiLi。展开 exampleDB 数据库,在"安全性"项下展开"用户"项,项中含有刚才建立的用户 LiLi。

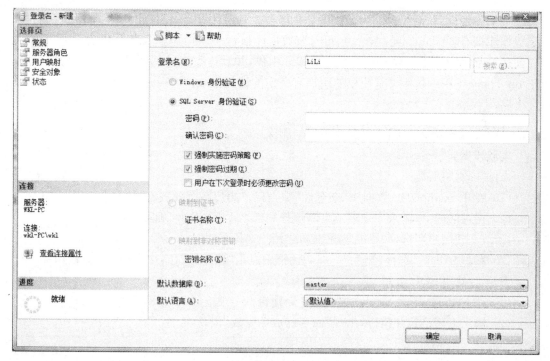

图 2-7 新建登录名对话框

方法 2：使用 SQL 语句

（1）创建新查询，输入下列语句：

```
USE master
CREATE LOGIN LiLi
WITH PASSWORD='123456',
DEFAULT_DATABASE=exampleDB
```

（2）执行查询语句。在"对象资源管理器"中，右击"登录名"，选择"刷新"，出现新登录名 LiLi。

（3）修改和删除登录名的语句分别是 ALTER LOGIN 和 DROP LOGIN。详细内容请查阅联机帮助。

2.3.3 用户与数据库角色

登录名连接至 SQL Server 服务器，服务器将针对这一登录名请求的数据库寻找对应的用户，用户获取相应授权并在权限范围内访问服务器上数据库。数据库的"安全性"项中，提供了专门管理用户与数据库角色的功能。

数据库创建后,系统自动为数据库添加了两个特殊用户 dbo 和 guest。dbo 用户的含义是数据库拥有者(database owner),可对该数据库执行一切操作。服务器角色 sysadmin 的任何成员都对应于每个数据库内的一个 dbo 用户,并且这类成员创建的所有对象都自动属于 dbo。guest 用户是让没有授权的用户有一定的访问权限。

【实验 2-10】 创建新用户与删除用户。

实验步骤:

方法 1:使用工具

(1) 展开 exampleDB 数据库"安全性"项,右击"用户",选择"新建用户"。

(2) 在对话框中,输入用户名 Alice,选择登录名 LiLi,点击"确定"按钮。

(3) 右击"用户",选择"刷新",显现新建用户 Alice。

(4) 右击"Alice",选择"删除"。在对话框中点击"确定"按钮,用户 Alice 被删除。

方法 2:使用 SQL 语句

(1) 在 Management Studio 中,点击"新建查询",建立查询编辑窗口。

(2) 输入下列语句并执行,出现"命令已成功完成"消息。

```
USE exampleDB
CREATE USER Alice FOR LOGIN LiLi
    WITH DEFAULT_SCHEMA = db_owner
```

(3) 输入下列语句并执行将从数据库中删除新建用户 Alice。

```
DROP USER Alice
```

数据库角色指定了以特定权限访问数据库对象的组。一个用户可以隶属于多个角色,角色权限具有累加性。用户通过某个数据库角色获得的权限可添加到同一个用户或通过其他数据库角色获得的权限上。系统在创建数据库时已为数据库定义了几个固定的数据库角色,角色的名称和权限如表 2-2 所示。

表 2-2　固定数据库角色的权限

固定数据库角色名称	权　　限
db_accessadmin	可以为 Windows 登录帐户、Windows 组和 SQL Server 登录帐户添加或删除访问权限。
db_backupoperator	可以备份该数据库。
db_datareader	可以读取所有用户表中的所有数据。
db_datawriter	可以在所有用户表中添加、删除或更改数据。

第 2 章　SQL Server 2008 数据库应用基础

（续表）

固定数据库角色名称	权　限
db_ddladmin	可以在数据库中运行任何数据定义语言（DDL）命令。
db_denydatareader	不能读取数据库内用户表中的任何数据。
db_denydatawriter	不能添加、修改或删除数据库内用户表中的任何数据。
db_owner	可以执行数据库的所有配置和维护活动。
db_securityadmin	可以修改角色成员身份和管理权限。
Public	每个数据库用户都属于 public 数据库角色。当尚未对某个用户授予或拒绝对安全对象的特定权限时，则该用户将继承授予该安全对象的 public 角色的权限。

上表中的固定数据库角色不能更改和删除，与服务器角色不同数据库角色可以由管理员新增和删除。

【实验 2-11】　创建与删除数据库角色。

实验步骤：

方法 1：使用工具

（1）在 Management Studio 中，展开 exampleDB 数据库的"安全性"下的"角色"项，右击"数据库角色"，选择"新建数据库角色"。

（2）在对话框中输入角色名称 myRole，单击"添加"按钮，选择 Alice 为此角色的成员，所有者不选，默认值为 dbo。单击"确定"按钮，完成新增数据库角色操作。

（3）删除数据库角色 myRole。右击 myRole 角色，选择"属性"，在对话框中删除角色成员 Alice。

（4）右击 myRole 角色，选择"删除"，在对话框中单击"确定"按钮，完成删除操作。如果没有删除数据库角色中的角色成员，直接删除角色则出现错误信息。

方法 2：使用 SQL 语句

（1）在 Management Studio 中，选择"新建查询"，在查询窗口中输入下列语句并执行。其中 sp_addrolemember 为系统存储过程，为数据库角色添加角色成员。

```
USE exampleDB
CREATE ROLE myRole AUTHORIZATION dbo
exec sp_addrolemember 'myRole','Alice'
```

（2）删除新建数据库角色的语句如下，其中为存储过程。

```
USE exampleDB
```

```
exec sp_droprolemember "myRole",'Alice'
DROP ROLE myRole
```

【实验项目】

实验项目 2.9

2.9 (1)以自己姓名拼音为登录名在数据库服务器上建立登录名,为此登录名不选择任何服务器角色。以该登录名登录服务器,点击 SPJ 数据库,分析出错原因。

(2)为 SPJ 数据库增加用户 Bob,选择登录名为上题建立的登录名。在"对象资源管理器"中断开当前服务器连接。重新以自己的登录名登录,观察 SPJ 数据库的表中内容,分析原因。

(3)断开当前连接,以 sa 管理员身份登录,为 Bob 添加 db_owner 数据库角色成员身份。再以自己的登录名登录,观察 SPJ 数据库的表中内容。

2.4　存储过程与触发器

2.4.1　存储过程

存储过程被称为存储在数据库中的程序,它由 SQL 语句和控制流语句构成。使用存储过程具有提高程序执行效率、易维护、减轻网络负担和安全等优点。初步了解存储过程并不难,但要深入掌握则有一定难度。

SQL Server 2008 在创建数据库时已生成一些系统存储过程,展开数据库下"可编程性"项的"存储过程"子项,双击"系统存储过程",可观察到许多系统存储过程。

【实验 2-12】 创建用户自定义的存储过程

实验步骤:

(1)在 exampleDB 数据库下,展开"可编程性",右击"存储过程",选择"新建存储过程",显现带有存储过程模板的查询文件窗口(其实用户自己新建查询文件也可以)。

(2)清除模板内容,输入下列程序(在模板中也可以)。

```
SET ANSI_NULLS ON
GO
SET QUOTED_IDENTIFIER ON
GO
```

第 2 章　SQL Server 2008 数据库应用基础

```
CREATE PROCEDURE  sp_bodyCount
    @dept varchar(20), ——系名
    @theBodyCount int OUTPUT ——已选课人数
AS
BEGIN
    SELECT @theBodyCount=count(DISTINCT SC.Sno)
    FROM Student,SC
    WHERE Student.Sno=SC.Sno and Sdept=@dept
END
GO
```

执行上面程序，右击"存储过程"，选择"刷新"，出现 dbo.sp_bodyCount 存储过程。存储过程的功能是输入系名，输出该系已选课的学生数。存储过程定义中引入了两个参数，第二个参数后面加了 OUTPUT，用来说明 @theBodyCount 参数的值可被调用者接收。

（3）建立新查询，在查询编辑窗口输入下列语句并执行，实现对存储过程的调用。

```
DECLARE @x int
exec sp_bodyCount 'CS',@x OUTPUT
print @x
```

2.4.2　触发器

触发器是一种特殊类型的存储过程，不由用户直接调用，数据表中数据被修改时执行。其功能是响应 INSERT、UPDATE 或 DELETE 语句的执行并产生相应操作。触发器可以查询其他表，并可以包含复杂的 Transact-SQL 语句。与 CHECK 约束不同，触发器可以引用其他数据表中的列，而且可以执行更为复杂的限制。

【实验 2-13】 在 exampleDB 数据库的 Student 数据表中触发器，用于检查学生年龄的正误，这里假设大学生的年龄不能小于 10 岁，也不能大于 30 岁。

实验步骤：

（1）在 Management Studio 的对象资源管理器中，展开 Student 表项，右击"触发器"，选择"新建触发器"，在查询编辑窗口中输入下列代码。

```
SET ANSI_NULLS ON
GO
SET QUOTED_IDENTIFIER ON
```

```
GO
CREATE TRIGGER tg_CheckStudentAge
   ON Student
   FOR INSERT,UPDATE
AS
BEGIN
    DECLARE @age int
    SELECT @age = inserted.Sage
    FROM inserted
    if @age<10 or @age>30
    BEGIN
        raiserror('输入错误',16,1,@age)
        ROLLBACK TRANSACTION
    END
END
GO
```

SQL Server 数据库管理系统自动创建并管理两个数据表 inserted 和 deleted，inserted 表存储了执行 INSERT 或 UPDATE 语句进行插入或更新数据表时被修改行的新记录，deleted 表存储着执行 DELETE 或 UPDATE 语句进行删除或更新数据表时被修改行的旧记录。程序员可以引用这两个数据表的信息，但不能修改。inserted 和 deleted 表的主要作用是供设计者获得数据表在修改前或修改后记录的值。

（2）执行查询窗口中的程序，点击对象资源管理器中的"触发器"，按 F5 键刷新，出现 tg_CheckStudentAge 触发器。

（3）新建查询，输入并执行下面语句，分析执行结果。

```
use exampleDB
go
INSERT into student
VALUES('200215256','李勇','男',9,'IS')
UPDATE Student
SET Sage=32
WHERE sno='200215123'
```

（4）将 student 表第一行记录的年龄改为 20，第二行记录的年龄改为 30。在查询中执行下列语句，并分析运行结果。修改 tg_CheckStudentAge 触发器，使其在处理多行记录时，只要有一行记录的年龄不满足要求，即终止修改。

update student
set sage=sage+1

【实验项目】

2.10 在 SPJ 数据库中完成下列设计：

（1）设计一个存储过程，要求输入工程项目代码，输出工程项目代码、工程项目名称、所使用的零件名称、零件重量与数量、零件的供应商名称。

（2）在 SPJ 表上创建一个触发器，当插入的 SNO、PNO 和 JNO 信息分别在 S 表、P 表和 J 表中不存在时，则放弃插入并反馈错误信息。

实验项目 2.10

2.5 数据库的维护

数据库在运行过程中，系统的硬件或软件故障、操作人员的失误等各种原因都会影响数据的正确性。出现故障后恢复数据的能力是衡量数据库管理系统性能的一项重要指标，SQL Server 2008 提供了强大的数据库维护工具，能方便地完成备份、恢复、镜像等工作。

2.5.1 分离和附加数据库

分离数据库是指将数据库从 SQL Server 实例中删除，但使数据库在其数据文件和事务日志文件中保持不变。之后，就可以使用这些文件将数据库附加到任何 SQL Server 实例，包括分离该数据库的服务器。

附加数据库是指将复制的或分离的数据库添加到 SQL Server 实例中。在 SQL Server 2008 中，数据库包含的全文文件随数据库一起附加。通常，附加数据库时会将数据库重置为它分离或复制时的状态。

【实验 2-14】 分离和附加 exampleDB 数据库。

实验步骤：

分离数据库的方法：

（1）在 SQL Server Management Studio 对象资源管理器中，展开"数据库"，右击数据库 exampleDB。

（2）选择"任务"，再单击"分离"。将显示"分离数据库"对话框。

(3)"选中要分离的数据库"网格将显示"数据库名称"列中选中 exampleDB 数据库。

(4)点击确定,exampleDB 数据库从对象资源管理器中消失,但在数据库系统所建立的文件夹的 Data 子文件夹下有文件 eaxmpleDB.mdf 和 example_log.ldf,数据库文件没有删除。

附加数据库的方法:

(1)在 SQL Server Management Studio 对象资源管理器中,右击"数据库",选择"附加"功能。将显示"附加数据库"对话框。

(2)在要附加的数据库下点击"添加"按钮,在"定位数据库文件"对话框中选择 exampleDB.mdf 文件。点击"确定"按钮,完成 exampleDB 的数据库附加。

2.5.2 脱机和联机数据库

数据库总是处于一个特定的状态中,脱机(离线)状态表示数据库无法使用。联机(在线)状态表示可以对数据库进行访问。

【实验 2-15】 脱机或联机 exampleDB 数据库。

实验步骤:

(1)在 SQL Server Management Studio 对象资源管理器中,脱机 exampleDB 数据库。右击 exampleDB 数据库,在弹出的快捷菜单中选择"任务",再选择"脱机",出现"使数据库脱机"对话框。

(2)联机 exampleDB 数据库。右击 exampleDB 数据库,在弹出菜单中选择"任务",再选择"联机",出现"使数据库联机"对话框。

2.5.3 备份与还原数据库

备份和还原数据库是数据库管理员的一项重要工作。SQL Server 提供了在不中断前台数据库工作的情况下执行备份和还原的能力。

备份分为三类:数据备份、差异备份和事务日志备份。

数据备份是指包含一个或多个数据文件的完整映像的任何备份。数据备份会备份所有数据和足够的日志,以便恢复数据。可对全部或部分数据库、一个或多个文件进行数据备份。

差异备份是基于之前进行的数据备份,称为差异的"基准备份"。每种主要的数据备份类型都有相应的差异备份。基准备份是差异备份所对应的最近完整或部分备份。差异备份仅包含基准备份之后更改的数据区。在还原差异备份之前,必须先还原其基准备份。

事务日志备份(又称"日志备份")中包括了在前一个日志备份中没有备份的所有日志记录。只有在完整恢复模式和大容量日志恢复模式下才会有事务日志备份。

还原可以选择的三种模式:简单模式、完整模式和大容量日志模式。

简单恢复模式简略地记录大多数事务,所记录的信息只是为了确保在系统崩溃或还原数据备份之后数据库的一致性。由于旧的事务已提交,已不再需要其日志,因而日志将被截断。截断日志将删除备份和还原事务日志。这种简化是有代价的,在灾难事件中有丢失数据的可能。没有日志备份,数据库只可恢复到最近的数据备份时间。简单恢复模式并不适合生产系统,因为对生产系统而言,丢失最新的更改是无法接受的。在这种情况下,Microsoft 建议使用完整恢复模式。

完整恢复模式完整地记录了所有的事务,并保留所有的事务日志记录,直到将它们备份。在 SQL Server Enterprise Edition 中,完整恢复模式能使数据库恢复到故障时间点(假定在故障发生之后备份了日志尾部)。

大容量日志恢复模式简略地记录大多数大容量操作(例如,索引创建和大容量加载),完整地记录其他事务。大容量日志恢复提高大容量操作的性能,常用作完整恢复模式的补充。大容量日志恢复模式支持所有的恢复形式,但是有一些限制。

【实验 2-16】 备份与还原 exampleDB 数据库。

实验步骤:

备份数据库的方法:

(1) 在 SQL Server Management Studio 对象资源管理器中,展开"数据库",右击数据库 exampleDB。在弹出菜单中选择"任务",再选择"备份",出现"备份数据库"对话框。如图 2-8 所示。

(2) 通过对话框可选择设定备份类型(完整、差异、事务日志)、备份组件(数据库、文件和文件组)、备份集、目标等内容。在对话框的选择页中点击"选项",可设置覆盖媒体、可靠性、事务日志和磁带机。这里使用默认设置,点击"确定"按钮,完成数据库备份。

数据库的还原操作与备份相似。为能观察到还原后的情况,建议先在数据表中新添一些数据。

还原 exampleDB 数据库的方法:

(1) 在 SQL Server Management Studio 对象资源管理器中,展开"数据库",右击 exampleDB 数据库。在弹出菜单中选择"任务",选择"还原",再选"数据库",弹出"还原数据库"对话框。

(2) 在对话框的选择页中点击"选项",在还原选项下,选中"覆盖现有数据库"。点击

"确定"按钮,完成数据库还原。

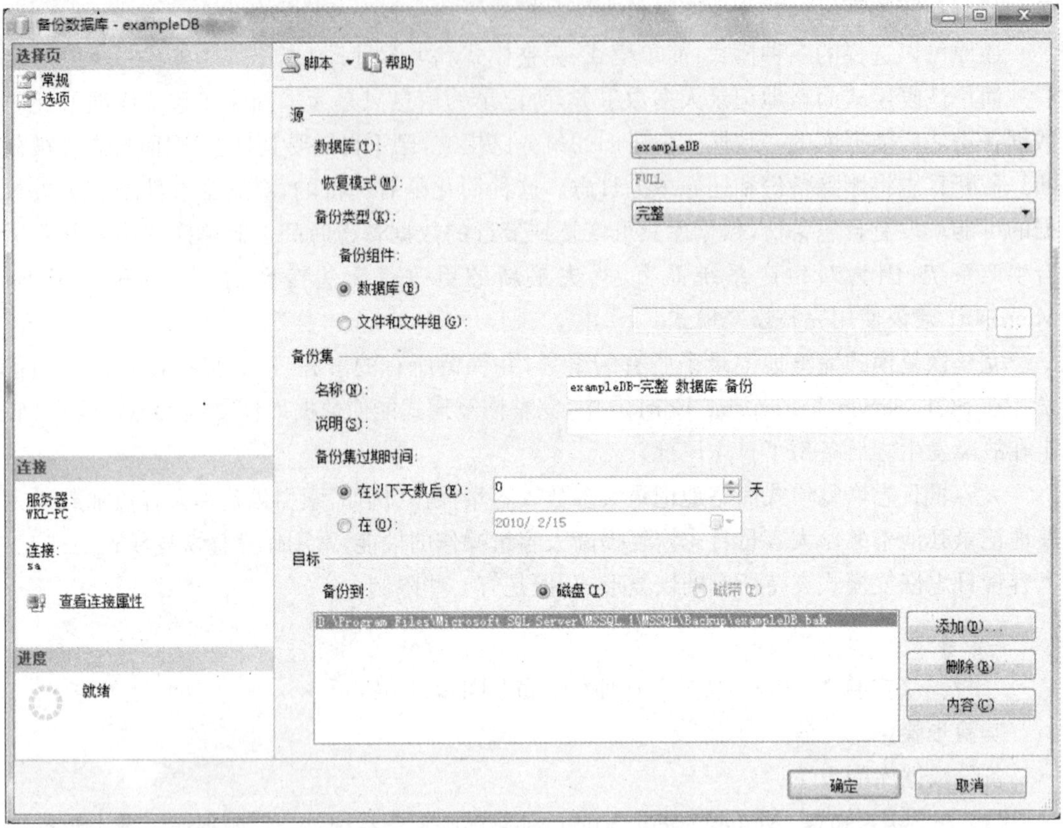

图 2-8 备份数据库对话框

2.5.4 导入与导出数据库

导入与导出是数据库管理的基本要求,其功能是在数据表和文件之间移动数据。SQL Server 允许用户大容量地导入和导出数据。导入是指将数据从数据文件加载到 SQL Server 表,导出是指将数据从 SQL Server 表复制到数据文件。SQL Server 提供了一组用于完成大容量导入和导出操作的通用工具和命令。

【实验 2-17】 Access 2003 中有一个联系人示例数据库(在帮助菜单下),将该数据库的内容导入至 SQL Server 数据库中。

实验步骤:

(1) 在 SQL Server Management Studio 对象资源管理器中,右击"数据库",在弹出的快捷菜单中选择"新建数据库"。在新建数据库对话框中输入新数据库名称 contact,点击"确定"按钮。

(2) 在对象资源管理器中,右击"contact"数据库,选择"任务",再选择"导入数据"。弹出"SQL Server 导入和导出向导"对话框,点击"下一步"进入"选择数据源"窗口。如图 2-9 所示。

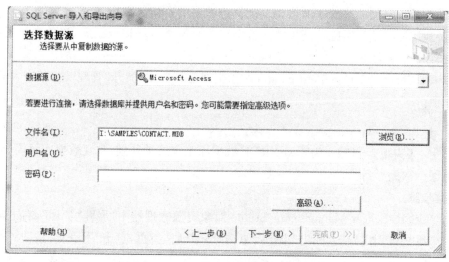

图 2-9 选择 Access 数据库为数据源窗口

(3) 在向导的"选择数据源"窗口中选择数据源为"Microsoft Access",点击"浏览"按钮选择 CONTACT.MDB 为文件名,点击"下一步"进入"选择目标"窗口。

(4) 在向导的"选择数据源"窗口中不改变任何选项,点击"下一步"进入"指定表复制或查询"窗口。

(5) 在向导的"指定表复制或查询"窗口中选择"复制一个或多个表或视图的数据"项(默认值),点击"下一步"进入"选择源表和源视图"窗口。

(6) 在向导的"选择源表和源视图"窗口中点击"全选"按钮,点击"下一步"进入"保存并执行包"窗口。

(7) 在向导的"保存并执行包"窗口中点击"下一步"进入"完成该向导"窗口。点击"完成"按钮,进入"执行"窗口,最后点击"关闭"按钮,完成导入。

【实验项目】

2.11 SQL Server 2008 提供了用备份与还原数据库的语句 BACKUP、RESTORE,数

据导入语句 BULK INSERT 和 INSERT ... SELECT * FROM OPENROWSET (BULK...)。查阅联机帮助,完成 exampleDB 数据库备份与还原,以及 Access 数据库数据导入到 SQL Server 数据库的实验。

2.5.5 制定维护计划

数据库中的数据是一种重要的资源,数据的损坏或丢失,可能带来无法弥补的损失。数据库管理员必须时刻关注的一项重要工作就是数据的备份。SQL Server 2008 提供了高性能的备份和还原功能。实施计划妥善的备份和还原策略可保护数据库,避免由于各种故障造成的损坏而丢失数据。

在 SQL Server Management Studio 对象资源管理器中,展开"管理"项,其中含有"维护计划"子项。

【实验 2-18】 为 exampleDB 数据库制定维护计划,实现数据库每周一次完整备份,每天一次差异备份。

实验步骤:

(1) 在 SQL Server Management Studio 对象资源管理器中,展开"管理",右击"维护计划",选择"维护计划向导"。弹出"维护计划向导"对话框首页,点击"下一步"按钮。

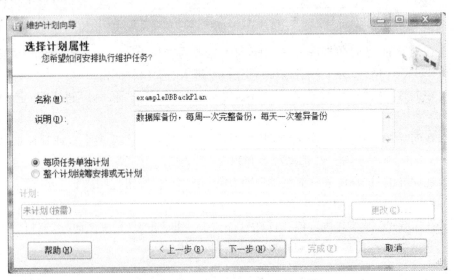

图 2-10 维护计划向导的选择计划属性页面

(2) 在维护计划向导的"选择计划属性"页,名称栏输入 exampleDBBackPlan 维护计划名称,说明栏输入一些解释性文字,并选择"每项任务单独计划",点击"下一步"按钮。

（3）在维护计划向导的"选择维护任务"页,点击"备份数据库(完整)"和"备份数据库(差异)"前的复选框,选中两项任务,点击"下一步"按钮。

（4）在维护计划向导的"选择维护任务顺序"页,直接点击"下一步"按钮。

（5）在维护计划向导的"定义"备份数据库(完整)"任务"页,选择数据库为 exampleDB,点击"更改"按钮,根据需求设置相应选项,点击"下一步"按钮。

（6）在维护计划向导的"定义"备份数据库(差异)"任务"页,完成与上一步相似的操作,点击"下一步"按钮。

（7）在维护计划向导的"选择报告选项"页,直接点击"下一步"按钮。

（8）在维护计划向导的"完成该向导"页,直接点击"完成"按钮。

（9）在 SQL Server Management Studio 对象资源管理器中,双击"exampleDBBackPlan"维护计划,可查阅维护计划详情。

【实验项目】

2.18 修改 exampleDBBackPlan 维护计划的"清除历史记录",设置如果保留时间超过 2 周,删除历史记录。类似地,尝试修改或设置维护计划中的内容。

实验项目 2.18

2.6 Reporting Services 应用

Reporting Services 是基于服务器的报表平台,支持各种数据源,可开发应用于窗体或 Web 应用程序的报表。它提供了诸多实用工具和服务,用于创建、部署和管理报表,并允许开发人员在自定义应用程序中集成或扩展数据和报表处理。平台所创建的报表可以通过基于 Web 的连接进行查看,也可以作为 Microsoft Windows 应用程序或 SharePoint 站点的一部分进行查看。

2.6.1 Reporting Services 配置管理器

SQL Server 2008 数据库管理系统安装过程中如果勾选了"Reporting Services"功能选择项,则在 SQL Server Management Studio 中展开"数据库"项,可见系统安装了"ReportServer"和"ReportServerTempDB"二个数据库。

Reporting Services 的配置通过系统提供的工具软件"Reporting Services 配置管理器"完成。工具软件启动后界面如图 2-11 所示。

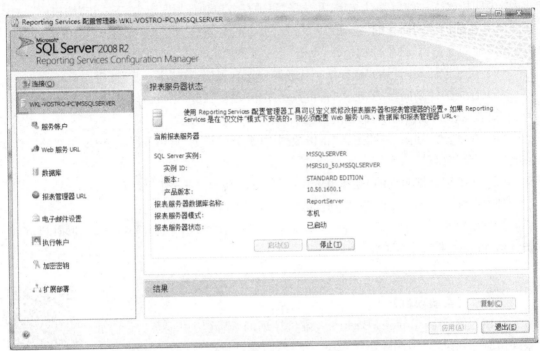

图 2-11 Reporting Services 配置管理器界面

　　Reporting Services 配置管理器的左侧为功能卡，点击 MSSQLSERVER 项，主界面显示报表服务器状态。如果"启动"按钮没有禁止，则点击该按钮使报表服务器为已启动状态。

　　"服务帐户"选项卡是设置运行该报表服务器的内置帐户或 Windows 域用户帐户。

　　"Web 服务 URL"选项卡是部署报表所在地址。报表服务器的默认虚拟目录为 ReportServer。

　　"数据库"选项卡是指新建设或更改报表服务器数据库。所有报表服务器内容和应用程序数据存储在该数据库中。

　　"报表管理器 URL"用于设置访问报表管理器的地址。远程用户可通过浏览器该地址登录报表管理器，其支持新建、修改和管理服务器中的报表。

　　"电子邮件设置"选项卡用于设置 SMTP 服务器和发件人地址。用户可通过电子邮件的方式分发及订阅报表。

　　"执行账户"、"加密密钥"和"扩展部署"三个选项卡的功能请查阅联机帮助。

　　完成报表服务器的配置后，验证报表服务器配置是否正确的方法是：在浏览器中，分别输入配置的报表管理 URL 和 Web 服务 URL，在弹出的用户验证窗口中输入正确的用户名和密码后，出现图 2-12 所示的报表管理器页面和图 2-13 所示的报表服务页面。

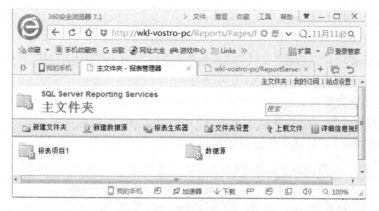

图 2-12 报表管理器页面

图 2-13 报表服务页面

2.6.2 报表设计

SQL Server 2008 报表的设计可通过 SQL Server Business Intelligence Development Studio 和 Report Builder 3.0 工具软件完成,其中前者已与 SQL Server 2008 软件包一起安装到数据库服务器中,后者则需要下载独立的软件包安装到用户的 PC 机中。

报表设计工具采用了与 Microsoft Office 相似的所见即所得的设计方式,并且在设计过程中可随时浏览报表的设计结果。

【实验 2-19】 设计一个基于 exampleDB 数据库的学生名单报表。

实验步骤:

(1) 打开 SQL Server Business Intelligence Development Studio 工具。点击"开始"按

钮,选择"所有程序"/"MicrosoftSQL Server 2008 R2"/"SQL Server Business Intelligence Development Studio"启动工具。软件运行界面参见图 2-14。

(2) 创建报表项目。点击"文件"/"新建"/"项目",弹出"新建项目"对话框,选择"报表服务器项目",输入名称:exampleDB 报表项目。

(3) 创建共享数据源。在解决方案资源管理器中,右击"共享数据源"项,从弹出菜单中选择"添加新数据源"。在"共享数据源属性"对话框中,输入名称:DS_exampleDB,点击"编辑"按钮,从"连接属性"对话框中选择连接字符串,点击"确定"。

图 2-14　exampleDB 报表项目设计窗口

(4) 创建报表。在解决方案资源管理器中,右击"报表"项,从弹出菜单中选择"添加新报表",弹出"报表制导"窗口。

点击"下一步",进入"选择数据源"页面,选择共享数据源为 DS_exampleDB。

点击"下一步",进入"设计查询"页面,点击"查询生成器",在"查询设计器"对话框的上半部分空白处,右击弹出菜单中选择"添加表",将 Student 表添加其中,并选择"所有列"项,点击"确定"。

点击"下一步",进入"选择报表类型"页面,选用缺省的"表格"项。

点击"下一步",进入"设计表"页面。点击"详细信息"按钮,将"可用字段"项中的所有字段选入"显示字段"项中。

点击"下一步",进入"选择表样式"页面,使用"石板"样式。

点击"下一步",进入"完成向导"页面,输入报表名称:StudentReport,点击"完成"。

在设计窗口,点击 StudentReport,Sno,Sname 等文本框,修改其中的英文为中文。点

第2章 SQL Server 2008 数据库应用基础

击"预览"页面,出现如图 2-15 所示界面。

图 2-15 报表项目预览窗口

2.6.3 报表部署

设计成功的报表需要发布到报表服务器方可供应用程序或用户直接访问。SQL Server Business Intelligence Development Studio 工具软件的部署功能可轻松实现报表的发布。报表部署前需要设置项目的 TargetServerURL 属性。

报表发布成功后,打开浏览器,输入在 Reporting Services 配置管理器中设置的 Web 服务 URL 链接,选择发布的报表项目,即可浏览到相应的报表。

【实验 2-20】 部署 exampleDB 报表项目到报表服务器并浏览报表。

实验步骤:

(1) 设置报表项目的 TargetServerURL 属性。打开 exampleDB 报表项目,从菜单栏目选择"项目"/"属性"(或者,在解决方案资源管理器中右击项目名称,选择"属性")。在 TargetServerURL 属性项中输入 http://localhost/ReportServer,点击"确定"。

(2) 部署报表。选择"生成"/"部署"(或者,在解决方案资源管理器中右击项目名称,选择"部署"),系统自动完成 exampleDB 报表到报表服务器的发布。

(3) 浏览已发布的报表。打开浏览器,在地址栏输入 http://localhost/ReportServer,弹出身份验证窗口,输入正确的用户信息。在浏览器中,点击"exampleDB 报表项目"链接,再点击"StudentReport"链接,浏览器中显示与图 2-15 相同的报表。

【实验项目】

2.19 设计课程成绩单报表,并部署其到报表服务器,通过浏览器查阅成绩单。

实验项目 2.19

第 3 章 数据库应用程序设计基础

Visual Studio 2010 软件开发平台为应用程序访问各种数据提供了一组功能丰富的类,被称为 ADO.NET,它是.NET Framework 中不可缺少的一部分。ADO.NET 提供对诸如 SQL Server 和 XML 这样的数据源以及通过 OLE DB 和 ODBC 公开的数据源的一致访问,应用程序可以使用 ADO.NET 连接到这些数据源,并可以检索、处理和更新其中包含的数据。

本章从最基础的连接开始,通过若干程序示例学习利用 ADO.NET 访问和处理数据库中数据的主要技术和实现方法。

3.1 应用程序与数据源连接的建立

3.1.1 ADO.NET 简介

ADO.NET 是微软对先前的数据访问技术 ADO(ActiveX Data Objects)一个跨时代的改进,它提供了平台互用性和可伸缩的数据访问。开发人员可利用它方便地对各种数据进行存取操作。

ADO.NET 用于访问和操作数据的两个主要组件是.NET Framework 数据提供程序和 DataSet。如图 3-1 所示。.NET Framework 数据提供程序是专门为数据操作以及快速、只进、只读访问数据而设计的组件。DataSet 是专门为独立于任何数据源的数据访问而设计的。因此,它可以用于多种不同的数据源,用于 XML 数据,或用于管理应用程序本地的数据。

访问数据的第一步是与数据源建立连接,.NET Framework 数据提供程序中的 Connection 对象专门负责与数据源的连接。Command 对象的作用是访问用于返回数据、修改数据、运行存储过程以及发送或检索参数信息的数据库命令。DataAdapter 对象在 DataSet 对象和数据源之间起到桥梁作用,DataAdapter 使用 Command 对象在数据源中执行 SQL 命令以向 DataSet 中加载数据,并将对 DataSet 中数据的更改解析回数据源。DataReader 可从数据源提供高性能的数据检索。

ADO.NET 与.NET Framework 中的 XML 类它们都是同一个体系结构的组件,ADO.NET 利用 XML 类的功能来提供对数据的断开连接方式的访问。无论 XML 源是文件还是 XML 流,都可以用其中的数据来填充 DataSet,DataSet 的本机序列化格式为 XML,因此它是用于在层间移动数据的最佳选择。

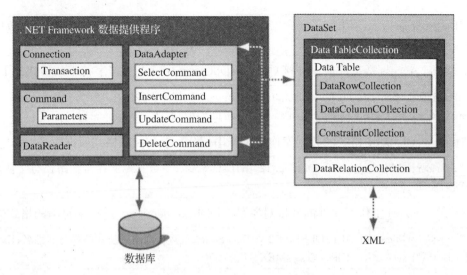

图 3-1 ADO.NET 结构框架图

3.1.2 连接数据源

.NET Framework 数据提供程序主要有下列四种：

(1) SQL Server.NET Framework 数据提供程序，提供对 Microsoft SQL Server 7.0 或更高版本中数据的访问。使用 System.Data.SqlClient 命名空间。

(2) OLE DB.NET Framework 数据提供程序，提供对使用 OLE DB 公开的数据源中数据的访问。使用 System.Data.OleDb 命名空间。

(3) ODBC.NET Framework 数据提供程序，提供对使用 ODBC 公开的数据源中数据的访问。使用 System.Data.Odbc 命名空间。

(4) Oracle.NET Framework 数据提供程序，提供对使用 Oracle 数据源中数据的访问。使用 System.Data.OracleClient 命名空间。

1. 访问 SQL Server 数据库通常用专门的数据提供程序，使用连接对象 SqlConnection，连接字符串可以用 Windows 身份验证或 SQL Server 身份验证。

Windows 身份验证是连接 SQL Server 的首选方法，连接字符的格式如下：

> "Persist Security Info=False;Integrated Security=true;Initial Catalog=数据库名称;Server=服务器"

"Persist Security Info=False;Integrated Security=SSPI;database=数据库名称;server=(local)"。其中，Persist Security Info=False 表示不允许在打开连接后通过连接获取安全敏感信息。

SQL Server 身份验证的连接字符串格式如下：

"Persist Security Info=False;User ID=用户登录名;Password=密码;Initial Catalog=数据库名称;Server=服务器名称"

2. 访问 Access 数据库使用连接对象 OleDbConnection。连接字符串的格式如下：

"Provider=Microsoft.Jet.OLEDB.4.0;Data Source=数据库文件所在路径与文件名;Jet OLEDB:System Database=d:\NorthwindSystem.mdw;User Id=用户名;Password=密码;"

对于 Access 2007 格式的数据库连接字符串（兼容 Access 旧格式数据库）的格式为：

"Provider=Microsoft.ACE.OLEDB.12.0;Data Source=数据库文件所在路径与文件名;Jet OLEDB:Database Password=密码;"

3. 访问 Excel 工作簿文件的连接对象是 OleDbConnection。连接字符串的格式如下：

"Provider=Microsoft.ACE.OLEDB.12.0;Data Source=数据库文件所在路径与文件名;Extended Properties=\"Excel 12.0;HDR=Yes;IMEX=1\""

其中，HDR=yes 是指第一行是列名而不是数据，IMEX=1 是指检索混合数据列是一种安全的方法。因 Extended Properties 的值有多项，必须用引号。

4. 访问 Oracle 数据库使用 OracleConnection 连接对象。连接字符串的格式如下：

"Data Source=Oracle9i;User ID=用户名;Password=密码;"

3.1.3 用配置文件保存连接字符串

连接数据库的字符串中一般会含有用户的登录名和密码，将它嵌入在应用程序代码中则可能成为安全漏洞，因为一些工具软件可以轻易地查看编译到应用程序源代码中的未加密连接字符串。此外，如果连接字符串发生更改，则必须重新编译应用程序。通常采用将连接字符串存储在应用程序配置文件中的方法保存连接字符串。

VS 2010 开发的应用程序可通过附加配置文件设置应用程序。在 Windows 应用程序可包含一个可选的 App.config 文件，在 ASP.NET 应用程序能包含一个或多个 Web.config 文件。System.Configuration 类封闭了访问与操作配置文件的方法，此外，系统还提供了加密保护配置文件中敏感信息的功能。

配置文件是 XML 格式的可编辑文件，实验 3-1 中 App.config 文件如下。

```
<? xml version="1.0" encoding="utf-8" ?>
<configuration>
  <connectionStrings>
    <add name="myDBConnection" connectionString="Persist Security Info=False;Integrated Security=SSPI;database=exampleDB;server=(local)"
      providerName="System.Data.SqlClient" />
```

```
        </connectionStrings>
    </configuration>
```

其中，<connectionStrings>的 name 项指定了程序连接数据库的字符串。在应用程序中，可通过如下两种方式获取连接字符串。

（1）用 Properties 读取

```
myConnString = Properties.Settings.Default.myDBConnection;
```

（2）用 ConfigurationManager 类读取

```
myConnString = System.Configuration.ConfigurationManager
              .ConnectionStrings["myDBConnection"].ConnectionString;
```

SQL Server 2008 支持使用 Windows 身份验证访问数据库，也是系统推荐的一种登录模式。这种模式不必再使用用户 ID 和密码，所以连接字符串中不含需要保护的信息。

对于配置文件中包含密码信息的处理，Visual Studio 2010 提供了对敏感信息加密保护功能，比较复杂，这里不再介绍，有兴趣的读者可参考联机帮助。

3.1.4 连接数据源示例程序

【实验 3-1】 编写窗体应用程序连接 SQL Server 数据库、Access 数据库和 Excel 文件，并列出其中的数据表名称。程序界面如图 3-2 所示。

实验步骤：

（1）创建 Windows 应用程序项目。拖拽选项卡控件 TabControl 于窗体，点击 TabPages 属性右侧按钮弹出 TabPage 集合编辑器对话框，添加 4 个 tabPage 成员，分别设置 Text 值为 SQL Server、Access、Excel、App.config。

（2）拖拽 TextBox 控件于窗体，设置 Multiline 属性为 True，改名称为 connMsgTxtBox。拖拽 OpenFileDialog 控件于窗体，改名称为 DBopenFileDialog。

（3）在 SQL Server 页面，拖拽 TextBox 控件 3 个，分别命名为 SQLDBNameTxtBox、SQLNameTxtBox、SQLPWDTxtBox，对应于数据库名称、登录名与密码的输入。再拖拽 Button 控件，命名为 btnConnSQLServer。类似地，在其他页面上根据需要拖拽控件这里不再赘述。

（4）添加 App.config 配置文件。在解决方案资源管理器中，右击项目，选择"添加"/"新建项"。在弹出的添加新项对话窗中，选择"应用程序配置文件"项，点击"添加"。

在 App.config 文件中添加数据库连接字符串如下：

```
<?xml version="1.0" encoding="utf-8"?>
<configuration>
```

```
<connectionStrings>
    <add name="myDBConnection"    connectionString="Persist Security Info
=False;Integrated Security=SSPI;database=exampleDB;server=(local)"
        providerName="System.Data.SqlClient" />
</connectionStrings>
</configuration>
```

（5）添加引用 System.configuration。在解决方案资源管理器中，展开项目下的"引用"项，查看有无"System.configuration"项。若没有 System.configuration，则右击"引用"，选择"添加引用"，弹出"添加引用"对话框，选择".NET"页，添加 System.configuration 项。

图 3-2　数据连接程序界面

程序代码：

```
private void btnConnSQLServer_Click(object sender, EventArgs e)
{    //连接 SQL Server 数据库
    string myConnString = "Data Source=(local);Initial Catalog=" + SQLDBNameTxtBox.Text
        +((SQLNameTxtBox.Text=="")?";Integrated Security = SSPI;":( ";User ID="+
SQLNameTxtBox.Text+";Password="+SQLPWDTxtBox.Text));
    var myConnection = new System.Data.SqlClient.SqlConnection(myConnString);
    try{
        myConnection.Open();        //打开连接
        ShowConnectionMsg(myConnection);   //显示连接状态信息
    }catch(SqlException ex) //处理异常
```

```
            {
                connMsgTxtBox.Text = "连接出错:" + ex.Message + Environment.NewLine;
            }finally
            {
                myConnection.Close();      //关闭连接
            }
        }

        private void AccessFileBtn_Click(object sender, EventArgs e)
        {   //打开文件浏览对话框选择 Access 数据库文件
            DBopenFileDialog.Filter = "Access 数据库|*.mdb|Access 2007 数据库|*.accdb";
            DBopenFileDialog.ShowDialog();
            AccessNametxtBox.Text = DBopenFileDialog.FileName;
        }

        private void AccessConnBtn_Click(object sender, EventArgs e)
        {   //连接 Access 数据库
            string myConnString = "Provider=Microsoft.ACE.OLEDB.12.0;" + "Data Source="
                + AccessNametxtBox.Text + ";Jet OLEDB:Database Password=" + AccessPwdtxtBox.Text;
            OleDbConnection myConnection = new OleDbConnection(myConnString);
            try{
                myConnection.Open();
                ShowConnectionMsg(myConnection);
            }catch (OleDbException ex)
            {
                connMsgTxtBox.Text = "连接出错:" + ex.Message + Environment.NewLine;
            }finally
            {
                myConnection.Close();
            }
        }

        private void ExcelFileBtn_Click(object sender, EventArgs e)
        {   //打开 Excel 文件
            DBopenFileDialog.Filter = "Excel 文件|*.xls|Excel 2007 文件|*.xlsx";
            DBopenFileDialog.ShowDialog();
            ExcelFileNameTxtBox.Text = DBopenFileDialog.FileName;
```

```csharp
}

private void ExcelConnBtn_Click(object sender, EventArgs e)
{
    //连接 Excel 文件
    string myConnString = "Provider=Microsoft.ACE.OLEDB.12.0;" + "Data Source="
        + ExcelFileNameTxtBox.Text + ";Extended Properties=\"Excel 12.0;HDR=Yes;IMEX=1\"";
    OleDbConnection myConnection = new OleDbConnection(myConnString);
    try{
        myConnection.Open();
        ShowConnectionMsg(myConnection);
    }catch (OleDbException ex)
    {
        connMsgTxtBox.Text = "连接出错:" + ex.Message + Environment.NewLine;
    }
    finally
    {
        myConnection.Close();
    }
}

private void AppConnBtn_Click(object sender, EventArgs e)
{
    //用 App.config 连接字符串连接 SQL Server 数据库,两种方法
    string myConnString=null;
    //方法1
    myConnString = Properties.Settings.Default.myDBConnection;
    //方法2,需要导入 System.Configuration.dll
    //myConnString = ConfigurationManager.ConnectionStrings["myDBConnection"].ConnectionString;
    var myConnection = new System.Data.SqlClient.SqlConnection(myConnString);
    try
    {
        myConnection.Open();
        ShowConnectionMsg(myConnection);
    }
    catch (SqlException ex)
    {
        connMsgTxtBox.Text = "连接出错:" + ex.Message + Environment.NewLine;
    }
```

```
        finally
        {
            myConnection.Close();
        }
    }
    private void ShowConnectionMsg(System.Data.Common.DbConnection myConnection)
    {   //在connMsgTxtBox控件显示连接信息和数据表名称
        connMsgTxtBox.Text = "***连接信息***\r\n连接状态:" + myConnection.State + Environment.NewLine;
        connMsgTxtBox.Text += "连接字符串:" + myConnection.ConnectionString + Environment.NewLine;
        connMsgTxtBox.Text += "数据源:" + myConnection.DataSource + Environment.NewLine;
        connMsgTxtBox.Text += "服务器版本:" + myConnection.ServerVersion + Environment.NewLine;
        connMsgTxtBox.Text += "数据库:" + myConnection.Database + Environment.NewLine;
        connMsgTxtBox.Text += "连接出错等待时间:" + myConnection.ConnectionTimeout + Environment.NewLine;
        //获取数据库中数据表名称
        connMsgTxtBox.Text += "\r\n***" + myConnection.Database + "数据源中含有的表***\r\n";
        DataTable dt = myConnection.GetSchema("Tables");
        foreach (System.Data.DataRow row in dt.Rows)
            connMsgTxtBox.Text += "数据表:" + row["TABLE_NAME"] + Environment.NewLine;
    }
```

3.2 用控件显示数据

3.2.1 DataSet、DataTable 和 DataAdapter 对象

DataSet 对象对于支持 ADO.NET 中的断开连接的分布式数据方案起到至关重要的作用。它可以用于多种不同的数据源,用于 XML 数据,或用于管理应用程序本地的数据。DataSet 表示包括相关表、约束和表间关系在内的整个数据集。

DataSet 是数据的一种内存驻留表示形式,无论数据来自什么数据源,都会提供一致的关系编程模型。DataSet 类包含了 DataTable 实例的集合,DataTable 实例包含了保存在数据集中的关系数据。每个 DataTable 实例包含 DataColumn 实例的集合用来定义表中的数据的架构,还包含 DataRow 实例的集合用来以数据行方式访问其中包含的数据。

DataTable 是.NET Framework 类库中 System.Data 命名空间的成员，是数据表在内存中的表示。程序员可以独立创建和使用 DataTable，也可以作为 DataSet 的成员创建和使用，而且 DataTable 对象也可以与其他.NET Framework 对象（包括 DataView）一起使用。还可以通过 DataSet 对象的 Tables 属性来访问 DataSet 中表的集合。

DataAdapter 用于从数据源检索数据并填充 DataSet 中的表，还将 DataSet 所做的更改解析回数据源，担负数据源数据和内存中缓存同步工作。DataAdapter 使用.NET Framework 数据提供程序的 Connection 对象连接到数据源，并使用 Command 对象从数据源检索数据以及将更改解析回数据源。

针对不同的连接对象，.NET Framework 数据提供程序对应地提供不同的 DataAdapter 对象。它们分别是适用于 OLE DB 的 OleDbDataAdapter 对象，适用于 SQL Server 的 SqlDataAdapter 对象，适用于 ODBC 的 OdbcDataAdapter 对象，适用于 Oracle 的 OracleDataAdapter 对象。

3.2.2 表中数据读至 DataSet 对象

在与数据源建立连接之后，通过 DataAdapter 将数据源中的数据表填充至 DataSet。DataAdapter 的 SelectCommand 属性是一个 Command 对象，用于从数据源中检索数据。DataAdapter 的 InsertCommand、UpdateCommand 和 DeleteCommand 属性也是 Command 对象，用于按照对 DataSet 中数据的修改来管理对数据源中数据的更新。

DataAdapter 的 Fill 方法用于使用 DataAdapter 的 SelectCommand 的结果来填充 DataSet。Fill 将要填充的 DataSet 和 DataTable 对象（或要使用从 SelectCommand 中返回的行来填充的 DataTable 的名称）作为它的参数。ADO.NET 访问数据源中关系表的主要步骤如下：

（1）建立连接

在前一节中，已介绍了各种连接方法。

（2）创建 DataAdapter 对象

```
SqlDataAdapter myAdapter = new SqlDataAdapter(SELECT 语句,连接字符串);
```

（3）创建 DataSet 对象

```
DataSet myDS = new DataSet();
```

（4）填充数据集

```
myAdapter.Fill(myDS,关系表名);
```

如果 DataAdapter 遇到多个结果集，将在 DataSet 中创建多个表。以 Table0 表示为第一个表名，Table1 为第二个，依次递推。如果以参数形式向 Fill 方法传递表名，则将向这些表提供递增的默认名称 TableNameN，以 TableName0 为"TableName"的第一个表名。可

第 3 章　数据库应用程序设计基础

以将任意数量的 DataAdapter 对象与 DataSet 一起使用。每个 DataAdapter 都可用于填充一个或多个 DataTable 对象并将更新返回相关数据源。

在数据填充时,Fill 方法发现连接尚未打开,它将隐式地打开 DataAdapter 正在使用的连接。如果 Fill 已打开连接,它还将在 Fill 完成时关闭连接。当处理单一操作(如 Fill 或 Update)时,这种方式可以简化代码。但是,如果执行多项需要打开连接的操作,则可以通过以下方式提高应用程序的性能:显式调用连接的 Open 方法,对数据源执行操作,然后调用连接的 Close 方法,释放链接。

注意:应保持与数据源的连接打开的时间尽可能短,以便释放资源供其他客户端应用程序使用。

3.2.3　数据显示控件与数据表的绑定

数据读取到 DataSet 后,可以使用各种控件显示数据。常用的有 DataGridView、Label、ListBox、TreeView 等,这些控件都有一个 DataBinding 属性,该属性为控件获取数据绑定。数据绑定含义是程序运行时控件自动将其属性和数据源关联在一起。

Windows 窗体可以利用两种类型的数据绑定:简单绑定和复杂绑定。这两种类型具有不同的优点。简单数据绑定是将一个控件绑定到单个数据元素,如 TextBox 或 Label 之类的控件绑定。事实上,控件上的任何属性都可以绑定到数据库中的字段。Visual Studio 中对此功能提供了广泛的支持。复杂数据绑定是将一个控件绑定到多个数据元素(通常是数据库中的多个记录)的能力。复杂绑定又被称作基于列表的绑定。支持复杂绑定的控件有 DataGridView、ListBox 和 ComboBox 控件。

BindingSource 组件用于简化将控件绑定到基础数据源的过程,该组件既可以作为一个媒介,也可以作为一个数据源,供其他控件绑定到该数据源。BindingSource 作为窗体上部分或全部组件的数据源,可以通过控件的 DataBindings 属性将 BindingSource 绑定到控件。

3.2.4　用控件显示数据示例程序

【实验 3-2】　连接 exampleDB 数据库,用 Label、Button、ListBox、DataGridView 等窗体控件显示关系表中数据。如图 3-3 所示。

实验步骤:

(1) 新建 Windows 窗体应用程序,拖拽两个 Button 控件,修改 Text 属性分别为"Student 数据表"和"三表连接",name 属性分别为"DataTableBtn"和"StuSCCourBtn"。拖拽 1 个 Label 控件、1 个 Button 控件、1 个 ComboBox 控件、1 个 ListBox 控件和 2 个 DataGridView 控件于窗体。

(2) 双击窗体,在 Form1_Load 函数中加入连接数据库代码。详见本节程序代码部分。

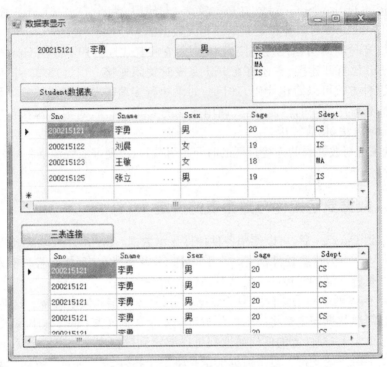

图3-3 利用窗体控件显示关系表

(3) 双击 DataTableBtn 按钮,在 DataTableBtn_Click 函数中加入填充 DataSet 数据集和数据绑定代码。

(4) 双击 StuSCCourBtn 按钮,在 StuSCCourBtn_Click 函数中完成 Student、SC、Course 三个关系表连接并在 DataGridView 中显示的代码。

程序代码:

```
public partial class Form1 : Form
{
    public Form1()
    {
        InitializeComponent();
    }
    private SqlConnection myConnection = null;
    private DataSet myDS;
    private void Form1_Load(object sender, EventArgs e)
    {
        string myConnString = null;
```

```csharp
            //获取App.config文件中与数据库连接的字符串
            myConnString = ConfigurationManager.ConnectionStrings["myDBConnString"]
                        .ConnectionString;
            //建立连接并测试成功与否
            myConnection = new System.Data.SqlClient.SqlConnection(myConnString);
            try
            {
                myConnection.Open();
                myDS = new DataSet();
            }
            catch (SqlException ex)
            {
                MessageBox.Show("连接出错:" + ex.Message, "提示信息", MessageBoxButtons.OK);

            }
            finally
            {
                myConnection.Close();
            }
        }

        private void DataTableBtn_Click(object sender, EventArgs e)
        {
            SqlCommand myComm = new SqlCommand("SELECT * FROM Student", myConnection);
            SqlDataAdapter myDataAdapter = new SqlDataAdapter(myComm);
            myDataAdapter.Fill(myDS, "Student");//填充数据到内存数据表
            //显示控件与数据集绑定
            label1.DataBindings.Add("Text", myDS.Tables["Student"],
                myDS.Tables["Student"].Columns["Sno"].ColumnName);
            button1.DataBindings.Add("Text", myDS.Tables["Student"],
                myDS.Tables["Student"].Columns[2].ColumnName);
            comboBox1.DataSource = myDS.Tables["Student"];
            comboBox1.DisplayMember = myDS.Tables["Student"].Columns["Sname"].ColumnName;
            listBox1.DataSource = myDS.Tables["Student"];
            listBox1.DisplayMember = myDS.Tables["Student"].Columns["Sdept"].ColumnName;
            dataGridView1.DataSource = myDS.Tables["Student"];
```

```
        }
        private void StuSCCourBtn_Click(object sender, EventArgs e)
        { //连接查询三表,查询结果填充至 myDS 数据集的另一数据表
            SqlCommand myComm = new SqlCommand("SELECT * FROM Student S LEFT OUTER JOIN SC SC1 ON S. Sno=SC1. Sno,SC SC2 RIGHT OUTER JOIN Course ON SC2. Cno= Course. Cno", myConnection);
            SqlDataAdapter adapter = new SqlDataAdapter(myComm);
            myConnection. Open();
            adapter. Fill(myDS);
            dataGridView2. DataSource = myDS. Tables[1];
            myConnection. Close();
        }
    }
```

【实验项目】

3.1 设计一个商品采购数据库,输入部分数据。编写应用程序,用窗体控件显示表中各种数据。

3.3 数据的插入、修改与删除

3.3.1 用 ADO. NET 维护数据表

数据表中记录的插入、删除、修改是一项基本工作,通常可使用结构化查询语言中的 INSERT、DELETE 和 UPDATE 语句完成相应的操作。在程序设计中,利用 ADO. NET 可方便地对数据库服务器中关系表进行各种操作,主要步骤如下:

(1) 建立连接

```
SqlConnection conn = new SqlConnection("连接字符串");
```

(2) 创建 SQL 语句字符串

```
String sql = "SQL 语句";
```

(3) 建立 SQL 命令对象

```
SqlCommand cmd = new SqlCommand(sql, conn);
```

第3章 数据库应用程序设计基础

```
    cmd.Parameters.Add…          //为 SQL 语句添加参数
```

(4) 执行

```
try
{
    conn.Open();
    cmd.ExecuteNonQuery();
}
catch(System.Data.SqlClient.SqlException ex)
{
    //处理异常
}
finally
{
    conn.Close();
}
```

3.3.2 数据表维护示例程序

【实验 3-3】 编程对 exampleDB 数据库中的 Student 表实现记录的插入、删除与修改。如图 3-4 所示。

实验步骤：

（1）新建 Windows 窗体应用程序。拖拽 DataGridView 控件、TabControl 控件、BindingSource 组件于窗体。

（2）点击 dataGridView1 控件，从属性窗口点击 Columns 属性右侧的按钮，弹出编辑列对话框，选择 Column1 列，在绑定列属性中修改 Width 为"60"，DataPropertyName 为"Sno"，HeaderText 为"学号"。类似地，完成其余列绑定属性的修改。

（3）在 tabControl1 控件中，点击 TabPages 属性右侧的按钮，弹出 TabPage 集合编辑器对话框。设置三个页，并分别修改 Text 为插入、修改、删除。在各个页中根据需要拖拽 Label、TaxtBox、ComboBox 等控件用于输入与显示，Button 实现插入、修改和删除功能。

（4）程序中主要函数功能。DisplayDB 函数完成从数据源读取关系表并填充 DataSct 对象，设置表中当前项的索引。bindingSource1_PositionChanged 函数在 Position 的值更改时发生，用 bindingSource1.Postion 的值更新 BindingManagerBase 对象的 position，实现改变 DataGridView 中选定的行，其他显示控件的内容也同步更改的功能。3 个 button 按钮的 Click 事件响应函数分别实现了插入、修改和删除功能。

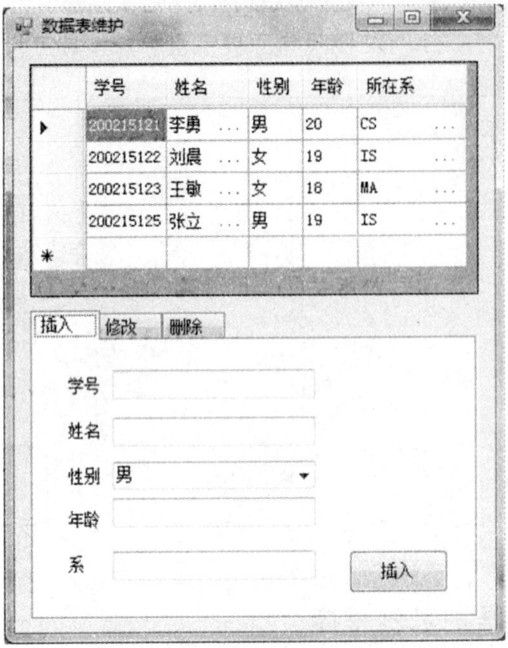

图 3-4 Student 数据表维护程序界面

程序代码:

```
public partial class Form1 : Form
{
    public Form1()
    {
        InitializeComponent();
    }

    private BindingManagerBase myBindingManagerBase;
    private DataSet ds = new DataSet();
    private SqlConnection Conn;
    private SqlDataAdapter adapter;

    private static string GetConnectionString()
    {
        return ConfigurationManager.ConnectionStrings["myDBConnString"].ConnectionString;
```

第 3 章　数据库应用程序设计基础

```csharp
}
private void DisplayDB()
{
    int pos = bindingSource1.Position;
    Conn = new SqlConnection(GetConnectionString());
    SqlCommand Comm = new SqlCommand("SELECT * FROM Student", Conn);
    adapter = new SqlDataAdapter(Comm);
    ds.Clear();

    Conn.Open();
    adapter.Fill(ds,"Student");
    myBindingManagerBase = BindingContext[ds, "Student"];
    myBindingManagerBase.Position = pos;

    bindingSource1.Position = pos;
    bindingSource1.DataSource = ds.Tables["Student"];
    dataGridView1.DataSource = bindingSource1;
    Conn.Close();
}

private void Form1_Load(object sender, EventArgs e)
{
    DisplayDB();
    textBox8.DataBindings.Add("Text", ds, "Student.Sno");    //绑定修改页中学号文本框
    textBox7.DataBindings.Add("Text", ds, "Student.Sname");  //绑定修改页中姓名文
                                                             本框
    textBox6.DataBindings.Add("Text", ds, "Student.Sage");   //绑定修改页中年龄文本框
    textBox5.DataBindings.Add("Text", ds, "Student.Sdept");  //绑定修改页中院系文本框
    label11.DataBindings.Add("Text", ds, "Student.Sno");     //绑定删除页中学号Label
    label12.DataBindings.Add("Text", ds, "Student.Sname");
    comboBox2.DataBindings.Add("Text", ds, "Student.Ssex");  //绑定修改页中性别组
                                                             合框
}

private void bindingSource1_PositionChanged(object sender, EventArgs e)
{   //点击dataGridView控件行,在修改和删除页中显示相同内容
```

```csharp
            myBindingManagerBase.Position = bindingSource1.Position;
        }

        private void button1_Click(object sender, EventArgs e)
        {   //插入新记录
            string InsertStr = string.Format("INSERT INTO Student VALUES('{0}','{1}','{2}','{3}','{4}')", textBox1.Text, textBox2.Text, comboBox1.Text, textBox3.Text, textBox4.Text);
            SqlCommand InsertComm = new SqlCommand(InsertStr, Conn);
            try
            {
                Conn.Open();
                InsertComm.ExecuteNonQuery();
            }
            catch (SqlException ex)
            {
                MessageBox.Show(ex.Message,"提示",
                                MessageBoxButtons.OK, MessageBoxIcon.Error);
            }
            finally
            {
                Conn.Close();
            }
            DisplayDB();
        }

        private void button2_Click(object sender, EventArgs e)
        {   //修改当前记录
            string UpdateStr = "UPDATE Student Set Sname=@Usname,Ssex=@Ussex,Sage=@Usage,Sdept=@Usdept" + " WHERE Sno=@Usno";

            SqlCommand UpdateComm = new SqlCommand(UpdateStr, Conn);

            UpdateComm.Parameters.Add(new SqlParameter("@Usno", SqlDbType.Char));
            UpdateComm.Parameters.Add(new SqlParameter("@Usname", SqlDbType.Char));
            UpdateComm.Parameters.Add(new SqlParameter("@Ussex", SqlDbType.Char));
            UpdateComm.Parameters.Add(new SqlParameter("@Usage", SqlDbType.Int));
```

```csharp
UpdateComm.Parameters.Add(new SqlParameter("@Usdept", SqlDbType.Char));

UpdateComm.Parameters["@Usno"].Value = textBox8.Text;
UpdateComm.Parameters["@Usname"].Value = textBox7.Text;
UpdateComm.Parameters["@Ussex"].Value = comboBox2.Text;
UpdateComm.Parameters["@Usage"].Value = textBox6.Text;
UpdateComm.Parameters["@Usdept"].Value = textBox5.Text;

try
{
    Conn.Open();
    UpdateComm.ExecuteNonQuery();
}
catch (SqlException ex)
{
    MessageBox.Show(ex.Message, "提示",
                    MessageBoxButtons.OK, MessageBoxIcon.Error);
}
finally
{
    Conn.Close();
}
DisplayDB();
}

private void button3_Click(object sender, EventArgs e)
{   //删除当前记录
    string DeleteStr = "DELETE FROM Student WHERE Sno='" + label11.Text.Trim() + "'";

    SqlCommand deleComm = new SqlCommand(DeleteStr, Conn);

    try
    {
        Conn.Open();
        deleComm.ExecuteNonQuery();
    }
    catch (SqlException ex)
```

```
            {
                MessageBox.Show(ex.Message,"提示",MessageBoxButtons.OK,
                        MessageBoxIcon.Error);
            }
            finally
            {
                Conn.Close();
            }
            DisplayDB();
        }
    }
```

【实验项目】

3.2　在实验 3.1 的基础上,设计具备数据库录入、修改和删除功能应用程序。

实验项目 3.2

3.4　调用数据库存储过程

3.4.1　ADO.NET 调用存储过程

存储过程可看成是在数据库服务器运行的程序,数据库应用程序利用存储过程实现各种功能是一种常见的设计方法。ADO.NET 提供了简便地调用存储过程的方法,主要步骤如下:

(1) 建立与数据库的连接。
(2) 创建调用存储过程的命令对象,指定存储过程名称。

```
SqlCommand cmd = new SqlCommand("numberStudentStoredProcedure", myConn);
```

(3) 说明命令类型是存储过程

```
cmd.CommandType = CommandType.StoredProcedure;
```

(4) 设置存储过程的输入与输出参数,输出参数需指明方向是 Output。

```
cmd.Parameters.Clear();
cmd.Parameters.Add("@CourseID", SqlDbType.Char).Value = Str;
cmd.Parameters.Add("@numberofStudent", SqlDbType.Int).Value = 0;
```

第3章 数据库应用程序设计基础

```
cmd.Parameters["@numberofStudent"].Direction = ParameterDirection.Output;
```

（5）打开连接，执行调用存储过程的命令，获取返回值，关闭连接。

```
try
{
    myConn.Open();
    cmd.ExecuteNonQuery();
    label2.Text = "选课人数:" + cmd.Parameters["@numberofStudent"].Value;
}
catch (SqlException ex)
{
    MessageBox.Show(ex.Message,"调用存储过程出错信息");
}
finally
{
    myConn.Close();
}
```

3.4.2 调用存储过程示例程序

【实验3-4】 用存储过程统计 exampleDB 数据库中每门课程的选课人数。如图3-5所示。

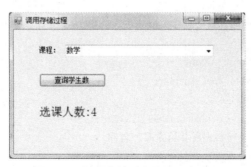

图 3-5 调用存储过程程序示例界面

实验步骤：

（1）打开 SQL Server Management Studio，在 exampleDB 数据库中添加存储过程。如果在 Visual Studio 2010 的服务器资源管理器中已建立与数据库的连接，同样也可创建存储过程。

```
CREATE PROCEDURE dbo.numberStudentStoredProcedure
(
    @CourseID char(10),
    @numberofStudent int OUTPUT
)
AS
    SELECT @numberofStudent=count(*) FROM SC WHERE Cno=@CourseID
    RETURN
```

（2）创建 Windows 窗体应用程序。从工具箱拖拽二个 Label 控件，修改 Text 属性分别为"课程："和"选课人数："。拖拽一个 ComboBox 控件和一个 Button 控件，修改 Button 控件的 Text 属性为"查询学生数"。

（3）双击 Form1 窗体，在 Form1_Load 事件响应函数中添加连接数据库，用课程名填充 comboBox1 的 DiaplayMember 属性的程序。详见程序代码部分。

（4）双击 button1 按钮，在 button1_Click 事件响应函数中添加调用存储过程的代码。详见程序代码部分。

程序代码：

```
public partial class Form1 : Form
{
    public Form1()
    {
        InitializeComponent();
    }
    private string connString = " Integrated Security = SSPI; Database = exampleDB; Server = localhost";
    private SqlConnection myConn;
    private SqlDataAdapter myDAdapter;
    private DataSet myDS;
    private void Form1_Load(object sender, EventArgs e)
    {   //程序加载时连接数据库并填充课程名称
        myConn = new SqlConnection(connString);
        myDAdapter = new SqlDataAdapter("SELECT * FROM Course", myConn);
        myDS = new DataSet();
        try
        {
            myConn.Open();
```

```
            myDAdapter.Fill(myDS,"Course");
            //用课程名填充 ComboBox 控件
            comboBox1.DataSource = myDS.Tables["Course"];
            comboBox1.DisplayMember = myDS.Tables["Course"].Columns["Cname"].ColumnName;
        }
        catch(SqlException ex)
        {
            MessageBox.Show(ex.Message,"访问数据库出错信息");
        }
        finally{
            myConn.Close();
        }
    }
    private void button1_Click(object sender, EventArgs e)
    {   //调用数据库中统计选课人数的存储过程
        DataRow[] CnoDR;
        string Str = "Cname = '" + comboBox1.Text+"'";
        CnoDR = myDS.Tables["Course"].Select(Str);
        Str=CnoDR[0][0].ToString();
        //利用 SqlDataAdapter 对象调用
        //myDAdapter.SelectCommand.CommandText = "numberStudentStoredProcedure";
        //myDAdapter.SelectCommand.CommandType = CommandType.StoredProcedure;
        //myDAdapter.SelectCommand.Parameters.Clear();
        // myDAdapter.SelectCommand.Parameters.Add("@CourseID", SqlDbType.Char).Value = Str;
        //myDAdapter.SelectCommand.Parameters.Add("@numberofStudent", SqlDbType.Int).Value = 0;
        // myDAdapter.SelectCommand.Parameters["@numberofStudent"].Direction = ParameterDirection.Output;
        //利用 SqlCommand 对象调用
        SqlCommand cmd = new SqlCommand("numberStudentStoredProcedure", myConn);
        cmd.CommandType = CommandType.StoredProcedure;
        cmd.Parameters.Clear();
        cmd.Parameters.Add("@CourseID", SqlDbType.Char).Value = Str;
        cmd.Parameters.Add("@numberofStudent", SqlDbType.Int).Value = 0;
        cmd.Parameters["@numberofStudent"].Direction = ParameterDirection.Output;
        try
```

```
            {
                myConn.Open();
                //myDAdapter.SelectCommand.ExecuteNonQuery();
                //label2.Text = "选课人数:" + myDAdapter.SelectCommand.Parameters["@numberofStudent"].Value;
                cmd.ExecuteNonQuery();
                label2.Text = "选课人数:" + cmd.Parameters["@numberofStudent"].Value;
            }
            catch (SqlException ex)
            {
                MessageBox.Show(ex.Message, "调用存储过程出错信息");
            }
            finally
            {
                myConn.Close();
            }
        }
    }
```

【实验项目】

3.3 在示例程序中,用存储过程实现添加统计每个系学生数的功能。

3.4 实验3-3程序中存在一个严重问题,即在删除学生记录前,没有先删除SC表中该生的选课信息。设计一个存储过程,实现删除学生记录同时删除选课记录,编程调用存储过程实现学生信息删除。

实验项目3.3 和3.4

3.5 数据库中图像的存取

3.5.1 图像存储的数据类型

数据库应用软件经常要求处理图片、声音、Word、动画等文件,在数据库中保存这些文档通常有二种方法。一种是在数据库中只保存文件所存储的路径,而文件内容不在关系表中存储。另一种是在数据表中设置专门字段并以二进制格式保存信息。

SQL Server 2008 提供了以二进制格式存储文件的能力,大值数据类型能存储最多可达 $2^{31}-1$ 字节的数据,可使用的大值数据类型有 varchar(max)、nvarchar(max) 和

varbinary(max)。较早的 SQL Server 使用 image 类型保存图像，在 SQL Server 2008 中建议用 varbinary(max)数据类型。

3.5.2 图像存取方法

图像的存取与字符、数值数据的存取主要过程基本一致，只是图像文件需要通过内存缓存方可实现存取。存入至数据表前需要先将文件读到内存缓存中，再用插入或更新 SQL 语句将缓存中信息保存到数据库。关键代码如下：

```
FileStream InFileStream = new FileStream(openFileDialog1.FileName, System.IO.FileMode.Open, System.IO.FileAccess.Read);
byte[] buffer = new byte[InFileStream.Length];
InFileStream.Read(buffer, 0, (int)InFileStream.Length);
SqlConnection Conn = new SqlConnection(getConnectionString());
string updateStr = string.Format("UPDATE Course SET CcoverPhoto=@photo WHERE Cname='{0}'",comboBox1.Text);
SqlCommand myCommand = new SqlCommand(updateStr, Conn);
myCommand.Parameters.Add("@photo", SqlDbType.VarBinary).Value = buffer;
Conn.Open();
myCommand.ExecuteNonQuery();
```

读取图像的过程是用 SELECT 语句从数据库中读出图像的二进制信息到内存缓存中，再用 PictureBox 控件显示。关键代码如下：

```
SqlConnection Conn = new SqlConnection(getConnectionString());
string seleStr = string.Format("SELECT CcoverPhoto FROM Course WHERE Cname='{0}'", comboBox1.Text);
SqlCommand myCommand = new SqlCommand(seleStr, Conn);
Conn.Open();
SqlDataReader dr = myCommand.ExecuteReader();
byte[] buffer = (byte[])dr[0];
System.IO.MemoryStream memStream = new System.IO.MemoryStream(buffer);
pictureBox1.Image = Image.FromStream(memStream);
```

3.5.3 数据库中图像存取示例程序

【实验3-5】在 exampleDB 数据库的 Course 表中增加教材封面字段，编程实现图像的导入与导出。如图 3-6 所示。

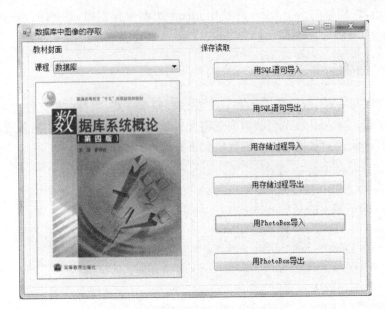

图 3-6 图像的存取示例程序界面

实验步骤：

（1）启动 SQL Server Management Studio，右击 Course 表，在弹出菜单中选择"设计"。在设计窗口添加新列，列名为"CcoverPhoto"，数据类型选择"varbinary(max)"，选中允许空项。

（2）在 exampleDB 数据库中创建存储过程。用于图像导入的存储过程如下：

```
CREATE PROCEDURE InputPhotoStoredProcedure
(
    @Cname char(50),
    @coverPhoto image
)
AS
    UPDATE  Course
    SET   CcoverPhoto = @coverPhoto
    WHERE   Cname = @Cname
    RETURN
```

用于图像导出的存储过程如下：

```
CREATE PROCEDURE OutputPhotoStoredProcedure
(
    @Cname char(50),
```

```
    @coverPhoto varbinary(MAX) OUTPUT
)
AS
    SELECT @coverPhoto=CcoverPhoto
    FROM Course
    WHERE Cname=@Cname
    RETURN
```

(3) 新建 Windows 窗体应用程序。拖拽两个 GroupBox 控件于 Form1 窗体,修改 Text 属性分别为"教材封面"和"保存读取"。拖拽 Label 控件至 groupBox1,设其 Text 属性为"课程"。拖拽 ComboBox 控件于 groupBox1,设其 DropDownStyle 属性为"DropDownList"。拖拽 PictureBox 控件于 groupBox1,设其 SizeMode 属性为"StretchImage",BorderStyle 属性为"FixedStyle"。在 groupBox2 控件中拖拽 6 个 Button 控件,依次设置 Text 属性内容。拖拽 OpenFileDialog 控件,设其 Filter 属性为"jpg 文件|*.jpg|bmp 文件|*.bmp"。

(4) 创建 App.config 文件,内容如下:

```
<?xml version="1.0" encoding="utf-8"?>
<configuration>
<connectionStrings>
<add name="myDBConnString" connectionString="Persist Security Info=False;Integrated Security=SSPI;database=exampleDB;server=(local)" providerName="System.Data.SqlClient"/>
</connectionStrings>
</configuration>
```

(5) 双击 Form1 窗体以及 6 个 Button 控件,在事件响应函数中添加代码,详见下面的程序代码部分。

程序代码:

```
public partial class Form1 : Form
{
    private SqlConnection conn;
    private SqlDataAdapter adapter;
    private DataTable table;

    public Form1()
    {
        InitializeComponent();
```

```csharp
        }
        private static string getConnectionString()
        {
            string connStr = null;
            ConnectionStringSettingsCollection mySettings = ConfigurationManager.ConnectionStrings;
            if (mySettings != null)
                foreach (ConnectionStringSettings setting in mySettings)
                    if (setting.Name == "myDBConnString")
                        connStr = setting.ConnectionString;
            return connStr;
        }

        private void Form1_Load(object sender, EventArgs e)
        {
            conn = new SqlConnection(getConnectionString());
            adapter = new SqlDataAdapter("SELECT * FROM Course", conn);
            table = new DataTable();
            adapter.Fill(table);
            comboBox1.DataSource = table;
            comboBox1.DisplayMember = table.Columns["Cname"].ColumnName;
        }

        private void button1_Click(object sender, EventArgs e)
        {   //用SQL语句插入图片至数据库
            if (openFileDialog1.ShowDialog() == DialogResult.OK)
            {
                FileStream InFileStream = new FileStream(openFileDialog1.FileName, System.IO.FileMode.Open, System.IO.FileAccess.Read);
                byte[] buffer = new byte[InFileStream.Length];//读入图片至内存缓存中
                InFileStream.Read(buffer, 0, (int)InFileStream.Length);
                //连接数据表,将图片更新至CcoverPhoto字段
                SqlConnection Conn = new SqlConnection(getConnectionString());
                string updateStr = string.Format("UPDATE Course SET CcoverPhoto=@photo WHERE Cname='{0}'", comboBox1.Text);
                SqlCommand myCommand = new SqlCommand(updateStr, Conn);
                myCommand.Parameters.Add("@photo", SqlDbType.VarBinary).Value = buffer;
                Conn.Open();
```

```
            try
            {
                myCommand.ExecuteNonQuery();
            }
            catch (SqlException ex)
            {
                MessageBox.Show(ex.Message,"插入图片出错");
            }
            Conn.Close();
            InFileStream.Close();
            buffer = null;
        }
    }

    private void button2_Click(object sender, EventArgs e)
    {   //用 SQL 语句导出至内存并显示
        using(SqlConnection Conn = new SqlConnection(getConnectionString()))//用于定义一个范围,在此范围的末尾将释放对象
        {
            string seleStr = string.Format("SELECT CcoverPhoto FROM Course WHERE Cname='{0}'", comboBox1.Text);
            SqlCommand myCommand = new SqlCommand(seleStr, Conn);
            Conn.Open();
            SqlDataReader dr = myCommand.ExecuteReader();
            if (dr.Read())
            {
                if (dr[0].GetType() != typeof(DBNull))
                {
                    byte[] buffer = (byte[])dr[0];
                    System.IO.MemoryStream memStream=new System.IO.MemoryStream(buffer);
                    pictureBox1.Image = Image.FromStream(memStream);
                }
                else
                    pictureBox1.Image = null;
            }
        }
```

```csharp
        }
        private void button3_Click(object sender, EventArgs e)
        {   //调用存储过程导入图片
            if (openFileDialog1.ShowDialog() == DialogResult.OK)
            {
                FileStream InFileStream = new FileStream(openFileDialog1.FileName, FileMode.Open, FileAccess.Read);
                byte[] buffer = new byte[InFileStream.Length];
                //读入图片至内存缓存中
                InFileStream.Read(buffer, 0, (int)InFileStream.Length);
                //连接数据库
                using (SqlConnection Conn = new SqlConnection(getConnectionString()))
                {
                    SqlCommand myCommand = new SqlCommand("InputPhotoStoredProcedure", Conn);
                    myCommand.CommandType = CommandType.StoredProcedure;
                    //命令为存储过程
                    //为存储过程添加参数
                    myCommand.Parameters.Add("@Cname", SqlDbType.Char).Value = comboBox1.Text;
                    myCommand.Parameters.Add("@coverPhoto", SqlDbType.VarBinary).Value = buffer;
                    try
                    {
                        Conn.Open();
                        myCommand.ExecuteNonQuery();
                    }
                    catch (SqlException ex)
                    {
                        MessageBox.Show(ex.Message, "用存储过程导入图片出错");
                    }
                }
                buffer = null;
            }
        }
        private void button4_Click(object sender, EventArgs e)
```

```csharp
    {
        //调用存储过程导出图片
        using (SqlConnection Conn = new SqlConnection(getConnectionString()))
        {
            SqlCommand myCommand = new SqlCommand("OutputPhotoStoredProcedure", Conn);
            myCommand.CommandType = CommandType.StoredProcedure;
            //命令为存储过程
            //为存储过程添加参数
            myCommand.Parameters.Add("@Cname", SqlDbType.Char).Value = comboBox1.Text;
            myCommand.Parameters.Add("@coverPhoto", SqlDbType.VarBinary, -1);
                                                                        //设置输出参数
            myCommand.Parameters["@coverPhoto"].Direction = ParameterDirection.Output;
            try
            {
                Conn.Open();
                myCommand.ExecuteNonQuery();
            }
            catch (SqlException ex)
            {
                MessageBox.Show(ex.Message, "用存储过程导出图片出错");
            }
            byte[] buffer = null;
            if (myCommand.Parameters["@coverPhoto"].Value.GetType() != typeof(DBNull))
            {
                buffer = (byte[])myCommand.Parameters["@coverPhoto"].Value;
                MemoryStream memStream = new MemoryStream(buffer);
                pictureBox1.Image = Image.FromStream(memStream);
            }
            else
                pictureBox1.Image = null;
        }
    }

    private void button5_Click(object sender, EventArgs e)
    {
        //利用 PictureBox 控件导入
        if (openFileDialog1.ShowDialog() == DialogResult.OK)
        {
```

```csharp
            using (SqlConnection Conn = new SqlConnection(getConnectionString()))
            {
                SqlDataAdapter adapter = new SqlDataAdapter("SELECT * FROM Course WHERE Cname='"+comboBox1.Text+"'",Conn);
                DataTable table = new DataTable();
                SqlCommandBuilder builder = new SqlCommandBuilder(adapter);
                adapter.Fill(table);
                BindingSource bindingSource = new BindingSource();
                bindingSource.DataSource = table;
                pictureBox1.DataBindings.Clear();
                pictureBox1.DataBindings.Add(new Binding("Image",bindingSource,"CcoverPhoto",true));
                if (pictureBox1.Image != null)
                    pictureBox1.Image.Dispose();
                pictureBox1.Image = Image.FromFile(openFileDialog1.FileName);
                bindingSource.EndEdit();
                adapter.Update(table);
            }
        }
    }

    private void button6_Click(object sender, EventArgs e)
    {  //利用PictureBox控件导出
        using (SqlConnection Conn = new SqlConnection(getConnectionString()))
        {
            SqlDataAdapter adapter = new SqlDataAdapter("SELECT * FROM Course WHERE Cname='" + comboBox1.Text + "'", Conn);
            DataTable table = new DataTable();
            adapter.Fill(table);
            byte[] buffer = null;
            if (table.Rows[0]["CcoverPhoto"].GetType() != typeof(DBNull))
            {
                buffer = (byte[])table.Rows[0]["CcoverPhoto"];
                pictureBox1.Image = Image.FromStream(new MemoryStream(buffer));
            }
            else
                pictureBox1.Image = null;
```

```
            }
        }
    }
```

【实验项目】

3.5 在 Student 表中添加用于存储学生照片信息的字段,编程实现照片的存取以及显示。

实验项目 3.5

3.6 主从关系数据表

3.6.1 主从数据表

在数据库应用程序中,主从数据表视图是一种非常常见的设计模式。在这种模式下,顶级数据(主表)显示在一个用户界面元素中(如 DataGridView),而与之相关的细节数据(从表)则根据主表的选择作相应的变化更新。

在.NET 中实现主从数据表模式一般有两种方式,一种是利用一个 DataGrid 控件装入两个相关的数据表,另一种是采用两个 DataGridView 控件实现主从模式。前一种方式由于主从表在同一个 DataGridView 控件中显示,操作时需要点击展开和返回按钮。后一种方式使用两个 DataGridView 控件,主从表均可见。对主表上某行进行选择,会立即引发从表内容的改变,直观性比前一种好。

3.6.2 主从数据表示例程序

【实验 3-6】 在 exampleDB 数据库中,设 Student 表为主表,从表内容选自 SC 和 Course 表。编程实现点击显示主表的 DataGridView 中行,立即在从表的 DataGridView 控件上显示该生相应的选课信息。如图 3-7 所示。

实验步骤:

(1) 新建 Windows 窗体应用程序。拖拽两个 GroupBox 控件于窗体,设置 Text 属性分别为"学生表"和"选课表"。拖拽两个 DataGridView 控件分别于 groupBox1 和 groupBox2 中,点击 dataGridView1 控件,从属性窗口点击 Columns 属性右侧的按钮,弹出编辑列对话框,点击"添加"5 次。选择 Column1 列,在绑定列属性中修改 Width 为"100", DataPropertyName 为"Sno",HeaderText 为"学号"。类似地,完成其余列绑定属性的设置。

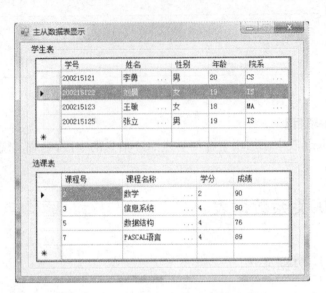

图 3-7 主从数据表显示示例程序界面

（2）在解决方案资源管理器中，右击项目，从弹出菜单中选择"添加"/"新建项"，在对话框中选择"应用程序配置文件"。在 App.config 文件中添加数据库连接字符串。展开"引用"，查看有无"System.configuration"项，若无，右击"引用"，选择"添加引用"，从添加引用对话框中选择 System.configuration 项。

（3）在 Form1.cs 文件前端添加 using System.Data.SqlClient；和 using System.Configuration；。双击 Form1 窗体，在 Form1_Load 函数中添加代码，内容详见下面的程序代码部分。选中 dataGridView1 控件，在属性窗口点击"事件"按钮，在 SelectionChanged 事件项的右侧双击，生成 dataGridView1_SelectionChanged 事件响应函数框架，向其中添加程序代码，代码内容详见下面的程序代码部分。

程序代码：

```
public partial class Form1 : Form
{
    private DataSet dataSet=new DataSet();
    public Form1()
    {
        InitializeComponent();
    }
    private static string getConnectionString()
    {
```

第3章 数据库应用程序设计基础

```
            return ConfigurationManager.ConnectionStrings["myDBConnString"]
.ConnectionString;
        }

        private void Form1_Load(object sender, EventArgs e)
        {
            //从 Student 表中获得学生信息产生 Table[0],从 SC 和 Course 表获取选课信息产生
Table[1]置于 dataSet 中
            var mySQLString = "SELECT * FROM Student;" + "SELECT SC.Sno,SC.Cno,
Course.Cname,Course.Ccredit,SC.Grade FROM SC,Course WHERE SC.Cno=Course.Cno";
            SqlDataAdapter myAdapter = new SqlDataAdapter(mySQLString,getConnectionString());
            myAdapter.Fill(dataSet);
            dataGridView1.DataSource = dataSet.Tables[0].DefaultView;
        }

        private void dataGridView1_SelectionChanged(object sender, EventArgs e)
        {
            //学生表的显示控件 dataGridView1 的当前选择项改变时执行
            //获取当前学号信息
            string myChoiceSno = dataSet.Tables[0].DefaultView[dataGridView1.CurrentRow!=null?
dataGridView1.CurrentRow.Index:0]["Sno"].ToString();
            //在第二个数据表中设置行过滤为 Sno=myChoiceSno
            dataSet.Tables[1].DefaultView.RowFilter = "Sno=" + myChoiceSno;
            dataGridView2.DataSource = dataSet.Tables[1];
        }
    }
```

【实验项目】

3.6 在第2章实验2.3所建数据库的基础上,以供应商表S为主表,以供应商供应的工程项目、零件、数量为从表。编程实现点击供应商信息,立即显示与该供应商相关的供货信息。

实验项目3.6

3.7 语言集成查询(LINQ)技术

3.7.1 LINQ 简介

语言集成查询(Language-Integrated Query)是微软公司提供的一项新技术。它能够

将查询功能直接引入到.NET Framework 3.5 所支持的编程语言(如 C♯、Visual Basic 等)中,使开发人员能够在应用程序代码中形成基于集合的查询,而不必使用单独的查询语言。查询操作可以通过编程语言自身来传达,而不是以字符串嵌入到应用程序代码中。

LINQ 技术主要包括 4 个独立技术:LINQ to Objects、LINQ to SQL、LINQ to DataSet 和 LINQ to XML。它们分别查询并处理对象数据(如集合等)、关系数据(如 SQL Server 数据库等)、DataSet 对象数据和 XML 结构(如 XML 文件)数据。使用 LINQ 可以大量减少了查询或操作数据库或数据源中的数据的代码,并在一定程度上避免了 SQL 注入,提供了应用程序的安全性。

在 LINQ to SQL 中,关系数据库的数据模型映射到用开发人员所用的编程语言表示的对象模型。当应用程序运行时,LINQ to SQL 会将对象模型中的语言集成查询转换为 SQL,然后将它们发送到数据库进行执行。当数据库返回结果时,LINQ to SQL 会将它们转换为可以被编程语言处理的对象。

3.7.2 对象关系设计器

在 Visual Studio 2008 中,若要实现 LINQ to SQL,首先需要用现有数据库的元数据创建 LINQ to SQL 对象模型。创建 LINQ to SQL 对象模型通常有下列三种方式:

(1) 对象关系设计器。该工具是 Visual Studio 2008 IDE 的一部分,支持可视化设计,适合中小型数据库。

(2) SQLMetal 命令行工具。对于大型数据库,这种命令行方式具有很好的可伸缩性,较多地用于大型数据库的建模。

(3) 代码编辑器。主要用于改进或修改由其他工具生成的代码,对象关系设计器(O/R 设计器)提供了一个可视化设计图面,用于将 LINQ to SQL 类映射到数据库中的表,这些映射的类称为"实体类"。开发者可用 O/R 设计器在应用程序中创建映射到数据库对象的对象模型,它生成一个强类型 DataContext,用于在实体类与数据库之间发送和接收数据。此外,O/R 设计器还提供了将存储过程和函数映射到 DataContext 方法以返回数据并填充实体类的功能。

O/R 设计器生成在 LINQ to SQL 类和数据库对象之间提供映射的.dbml 文件。O/R 设计器还生成类型化的 DataContext 和实体类。

O/R 设计器的设计图面有两个不同的区域:左侧的实体窗格以及右侧的方法窗格。实体窗格是主设计图面,其中显示实体类、关联和继承层次结构。方法窗格是显示映射到存储过程和函数的 DataContext 方法的设计图面。

3.7.3 查询表达式

查询表达式是用查询语法表示的查询。它就像任何其他表达式一样，并且可以用在 C♯ 表达式有效的任何上下文中。查询表达式由一组用类似于 SQL 或 XQuery 的声明性语法编写的子句组成。每个子句又包含一个或多个 C♯ 表达式，而这些表达式本身又可能是查询表达式或包含查询表达式。

查询表达式必须以 from 子句开头，并且必须以 select 或 group 子句结尾。在第一个 from 子句和最后一个 select 或 group 子句之间，查询表达式可以包含一个或多个下列可选子句：where、orderby、join、let 甚至附加的 from 子句。还可以使用 into 关键字使 join 或 group 子句的结果能够充当同一查询表达式中附加查询子句的源。

在 LINQ 中，查询变量是任何存储查询而不是查询结果的变量。更具体地说，查询变量始终是一个可枚举的类型，当在 foreach 语句中或对其 IEnumerator.MoveNext 方法的直接调用中循环访问它时，它将产生一个元素序列。例如：

```
int[] numbers = { 5, 4, 1, 3, 9, 8, 6, 7, 2, 0 };  //创建数据源 numbers
var lowNums = from num in numbers      // lowNums 为查询变量，num 为范围变量
    where num < 5
    select num;
foreach (int i in lowNums)
{
    Console.Write(i + " ");
}
```

查询表达式可以包含多个 from 子句。当源序列中的每个元素本身就是集合或包含集合时，可使用附加的 from 子句。例如：

```
var cityQuery =
    from country in countries
    from city in country.Cities
    where city.Population > 10000
    select city;
```

用 select 子句可产生所有其他类型的序列。简单的 select 子句只是产生与数据源中包含的对象具有相同类型的对象的序列。可以使用 select 子句将源数据转换为新类型的序列。这一转换也称为"投影"。例如：

```
var queryNameAndPop =
    from country in countries
```

```
select new { Name = country.Name, Pop = country.Population };
```

用 group 子句可产生按照指定的键组织的组序列。键可以采用任何数据类型。例如:

```
var queryCountryGroups =
    from country in countries
    group country by country.Name[0]; //键值为 country.Name[0],字符串类型
```

在 select 或 group 子句中使用 into 关键字来创建用于存储查询的临时标识符。当必须在分组或选择操作之后对查询执行附加查询操作时,需要这样做。例如:

```
var percentileQuery =
    from country in countries
    // let 子句可以将表达式的结果存储到新的范围变量中
    let percentile = (int) country.Population / 10000000
    //以一千万人口范围为界对 countries 进行分组
    group country by percentile into countryGroup
    where countryGroup.Key >= 20
    orderby countryGroup.Key
    select countryGroup;
```

筛选、排序和联接。在 from 开始子句以及 select 或 group 结束子句之间,所有其他子句(where、join、orderby、from、let)都是可选的。任何可选子句都可以在查询正文中使用零次或多次。使用 where 子句可以根据一个或多个谓词表达式筛选掉源数据中的某些元素。使用 orderby 子句可以按升序或降序对结果进行排序。使用 join 子句可以根据每个元素中指定键之间的相等比较,对一个数据源中的元素与另外一个数据源中的元素进行关联或组合。例如:

```
var myQuery = from sc in myDataContext.SC
    join student in myDataContext.Student on sc.Sno equals student.Sno
    where sc.Cno == comboBox1.SelectedValue.ToString()
    orderby sc.Sno ascending
    select new
    {
        学号=sc.Sno,
        姓名=student.Sname,
        成绩=sc.Grade,
        院系=student.Sdept
    };
```

3.7.4 LINQ to SQL 访问 SQL 数据库

使用 LINQ to SQL 技术访问 SQL Server 2008 数据库的典型步骤为：第一步是用现有关系数据库的元数据创建对象模型；第二步使用对象模型，在模型中描述信息请求和数据操作。

在 Visual Studio 2008 中，可以使用 O/R 设计器来创建对象模型，也可以使用命令行代码生成工具 SQLMetal。

使用已创建的对象模型。通过创建查询表达式以从数据库中检索信息，重定义 Insert、Update 和 Delete 的默认行为，实现数据表的插入、更新和删除操作。

LINQ to SQL 在操作和保持对对象所做更改方面有着最大的灵活性。实体对象可用后，就可以像应用程序中的典型对象一样更改实体对象。也就是说，可以更改它们的值，将它们添加到集合，以及从集合中移除它们。LINQ to SQL 会跟踪所做的更改，并且在调用 SubmitChanges 时就可以将这些更改传回数据库。

LINQ to SQL 不支持且无法识别级联删除操作。如果要在对行有约束的表中删除行，则必须在数据库的外键约束中设置 ON DELETE CASCADE 规则，或者使用自己的代码首先删除防止删除父对象的子对象。否则会引发异常。

3.7.5 LINQ 技术访问数据库示例程序

【实验 3-7】 在 exampleDB 数据库中，用 LINQ to SQL 技术编程实现具有以下功能的应用程序。如图 3-8 所示。

功能：1. 查阅每门课学生名单，2. 增添新课程，3. 删除课程和该课的选课信息。

图 3-8 LINQ 查询示例程序主界面

实验步骤：

(1) 新建 Windows 窗体应用程序。在解决方案资源管理器中，右击项目，选择"添加"/"新建项"。在弹出的"添加新项"对话框中，从"模板"中选择"LINQ to SQL 类"，用默认的名称 DataClasses1.dbml，点击"添加"按钮。出现 O/R 设计器主界面，如图 3-9 所示。

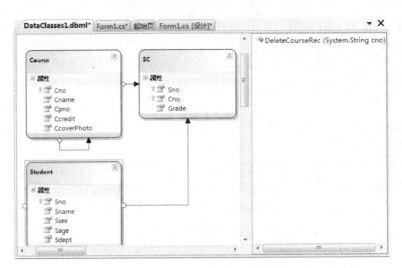

图 3-9 O/R 设计器界面

（2）在服务器资源管理器中，添加与 exampleDB 数据库的连接（如已建立可省略）。打开连接中的"表"项，拖拽 Student、SC、Course 三个表到 O/R 设计器。在 exampleDB 数据库中，新建用于 Course 记录删除的存储过程，如下：

```
CREATE PROCEDURE dbo.DeleteCourseRec
(
    @cno char(4)
)
AS
    DELETE FROM SC WHERE Cno=@cno
    DELETE FROM Course WHERE Cno=@Cno
    RETURN
```

拖拽存储过程到 O/R 设计器，存储过程出现的设计器右侧窗格，如图 3-9 所示。至此，完成了用 O/R 设计器创建对象模型的工作。

（3）在 Form1 窗体中播放 3 个 Button 控件，修改 Text 属性分别为"选课状况"、"添加课程"和"删除课程"。在解决方案资源管理器中，右击项目，选择"添加"/"Windows 窗体"，依次添加 3 个新窗体，名称分别为 Form2、Form3、Form4。3 个窗体分别用来在主窗体中点击"选课状况"、"添加课程"和"删除课程"按钮时产生相应对话框。

（4）在 Form2 窗体中，修改 Text 属性为"选课查询"。拖拽 1 个 Label 控件、1 个 ComboBox 控件和 1 个 DataGridView 控件于 Form2 窗体，并修改控件属性：lable1 的 Text 属性为"课程"、comboBox1 的 DisplayMember 属性为"Cname"、DropDownStyle 属性为

"DropDownList"、ValueMember 属性为"Cno"。

（5）在 Form3 窗体中，修改 Text 属性为"添加课程"。拖拽 4 个 Label 控件和 TextBox 控件、1 个 Button 控件和 1 个 DataGridView 控件于 Form3 窗体，并修改 DataGridView 控件的 Columns 属性，添加 4 个列用于显示课程编号（Cno）、课程名称（Cname）、先导课程（Cpno）和学分（Credit）。4 个 Lable 控件的 Text 属性分别为"课程编号"、"课程名称"、"先导课程"和"学分"，Button 控件的 Text 属性为"添加"。

（6）在 Form4 窗体中，修改 Text 属性为"删除课程"。拖拽 1 个 Label 控件、1 个 ComboBox 控件、1 个 Button 控件于 Form4 窗体。修改 Label 控件的 Text 属性为"课程"，修改 ComboBox 控件的 DisplayMember 属性为"Cname"、DropDownStyle 属性为"DropDownList"、ValueMember 属性为"Cno"，修改 Button 控件的 Text 属性为"调用存储过程删除课程"。

（7）在 Form1 窗体中，双击 button1 按钮，为 Click 事件添加显示 Form2 窗体的代码（Form2 form2 = new Form2(); form2.ShowDialog();)，详见程序代码部分。类似地完成 button2 和 button3 的 Click 事件响应函数。

（8）在 Form2.cs 文件中，添加私有函数 showDataGridView() 用于显示每门课的选课情况，并供 Form2_Load 和 comboBox1_SelectedValueChanged 函数调用。Form3.cs 和 Form4.cs 与 Form2.cs 类似，详见程序代码部分。注意：录入代码时事件响应函数的函数框架是通过双击相应事件产生的。

程序代码：

```
public partial class Form1 : Form
{   //主窗体的程序代码
    public Form1()
    {
        InitializeComponent();
    }
    private void button1_Click(object sender, EventArgs e)
    {   //创建选课查询子窗体对象,以对话框模式显示
        Form2 form2 = new Form2();
        form2.ShowDialog();
    }
    private void button2_Click(object sender, EventArgs e)
    {   //创建添加课程子窗体对象,以对话框模式显示
        Form3 form3 = new Form3();
        form3.ShowDialog();
```

```csharp
        }
        private void button3_Click(object sender, EventArgs e)
        {   //创建删除课程子窗体对象,以对话框模式显示
            Form4 form4 = new Form4();
            form4.ShowDialog();
        }
}

public partial class Form2 : Form
{          //选课查询子窗体的代码
    public Form2()
    {
        InitializeComponent();
    }
    private void showDataGridView()
    {
        DataClasses1DataContext myDataContext = new DataClasses1DataContext();
        var myQuery = from sc in myDataContext.SC
                      join student in myDataContext.Student on sc.Sno equals student.Sno
                      where sc.Cno == comboBox1.SelectedValue.ToString()
                      orderby sc.Sno ascending
                      select new
                      {
                          学号=sc.Sno,
                          姓名=student.Sname,
                          成绩=sc.Grade,
                          院系=student.Sdept
                      };
        this.dataGridView1.DataSource = myQuery;
    }
    private void Form2_Load(object sender, EventArgs e)
    {
        DataClasses1DataContext myDataContext = new DataClasses1DataContext();
        var myQuery = from course in myDataContext.Course
                      select course;
        //要先设置 comboBox 的 DisplayMember=Cname,ValueMember
```

```csharp
            this.comboBox1.DataSource = myQuery;
            showDataGridView();
        }
        private void comboBox1_SelectedValueChanged(object sender, EventArgs e)
        {
            showDataGridView();
        }
}

public partial class Form3 : Form
{       //添加课程子窗体的代码
        public Form3()
        {
            InitializeComponent();
        }
        private void showDataGridView()
        {
            DataClasses1DataContext myDataContext = new DataClasses1DataContext();
            var myInsertQuery = from course in myDataContext.Course
                                select new
                                {
                                    cno=course.Cno,
                                    name=course.Cname,
                                    pno=course.Cpno,
                                    credit=course.Ccredit
                                };
            this.dataGridView1.DataSource = myInsertQuery;
        }
        private void Form3_Load(object sender, EventArgs e)
        {
            showDataGridView();
        }
        private void button1_Click(object sender, EventArgs e)
        {       //用实体类方法插入一行数据
            DataClasses1DataContext myDC = new DataClasses1DataContext();
            Course course = new Course
            {
```

```csharp
                Cno=textBox1.Text,
                Cname=textBox2.Text,
                Cpno=(textBox3.Text==""? null:textBox3.Text),
                Ccredit=Convert.ToInt16(textBox4.Text)
            };
        myDC.Course.InsertOnSubmit(course);
        myDC.SubmitChanges();
        showDataGridView();
        }
}

public partial class Form4 : Form
{    //删除课程子窗体的代码
    public Form4()
    {
        InitializeComponent();
    }
    private void showComboBox()
    {
        DataClasses1DataContext myDataContext = new DataClasses1DataContext();
        var myQuery = from course in myDataContext.Course
                      select course;
        this.comboBox1.DataSource = myQuery;
    }
    private void Form4_Load(object sender, EventArgs e)
    {
        showComboBox();
    }
    private void button1_Click(object sender, EventArgs e)
    {
        DataClasses1DataContext myDC = new DataClasses1DataContext();
        myDC.DeleteCourseRec(comboBox1.SelectedValue.ToString());
        showComboBox();
    }
}
```

【实验项目】

3.7 在本节示例程序的基础上,用 LINQ to SQL 技术实现为每门课程增删学生的程序。

实验项目 3.7

3.8 服务器报表应用程序设计

3.8.1 ReportViewer 控件

ReportViewer 是 Visual Studio 2010 中用于显示报表及其关联功能的控件。该控件支持本地处理和远程处理两种工作模式。在本地处理模式中,控件在本地处理和显示报表。在远程处理模式中,报表在报表服务器上处理,在本地显示。远程处理模式需要报表服务器为 Microsoft SQL Server 2008 或更高版本实例。

在 Visual Studio 2010 中,无论是开发 Windows 窗体应用程序还是 ASP.NET Web 应用程序,均可方便地应用 ReportViewer 控件处理存储在本地或远程报表服务器报表。

ReportViewer 控件的 ProcessingMode 属性用于获取或设置控件的处理模式。在用枚举量 ProcessingMode.Remote 或 ProcessingMode.Local 进行设置时,窗体应用程序需要引用 Microsoft.Reporting.WinForms,而 Web 应用程序需要引用 Microsoft.Reporting.WebForms。

ReportViewer 控件的 ReportServerUrl 属性用于获取或设置报表服务器的 URL。

ReportViewer 控件的 ReportPath 属性用于获取或设置到报表服务器上报表的路径。

以上三个属性可通过 ReportViewer 控件的智能标记菜单或编写代码进行设置。

3.8.2 服务器报表示例程序

【实验 3-8】 设计学生选课成绩单报表,分别在窗体应用程序和 Web 应用程序中显示报表。如图 3-10 所示。

实验步骤:

(1) 在报表服务器中创建学生选课成绩单报表。启动 SQL Server Business Intelligence Development Studio 工具,打开在第 2 章实验 2-19 中创建的 exampleDB 报表项目。右击"报表"项,选择"添加"/"新建项"/"报表",名称为缺省的 Report1.rdl,点击"添加"按钮。

配置数据集。从工具箱拖拽"表"到设计窗口,在弹出的数据集属性对话窗口中,选择

"使用在我的报表中嵌入的数据集",点击"新建"按钮,选择"使用共享数据引用",下拉列表框中选择 DS_exampleDB 项(在实验 2-19 中已经创建此共享数据源),选择"确定"。点击"查询设计器"按钮,从弹出的查询设计器对话窗口中,在设计窗口右击,选择"添加表",依次选择 Student,SC 和 Course 数据表,并在 Student 和 Course 表选择"所有列"项,SC 表仅选 Grade 项,点击"确定"。

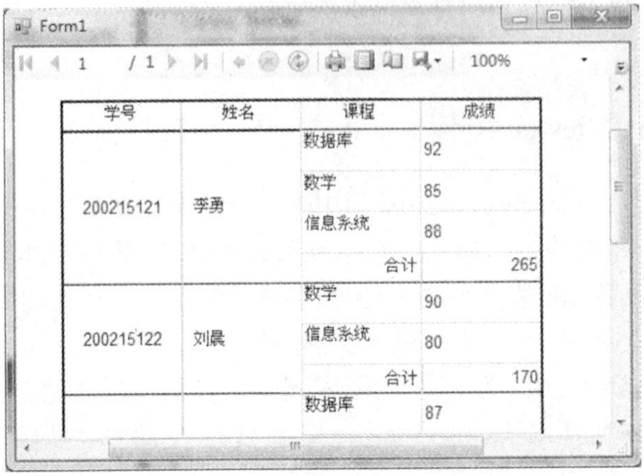

图 3-10 报表窗体应用程序中的显示界面

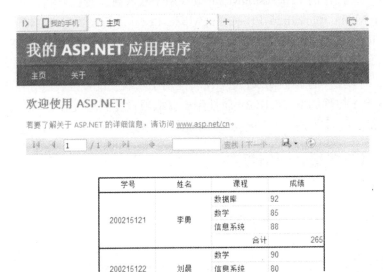

图 3-11 报表 Web 应用程序中的显示页面

设计报表。光标到 Tablix1 控件第 1 列第 2 行,从右上角弹出的菜单项中,选择"Cname"。类似地,在第 2 列第 2 行选择"Grade"。选中 Tablix1 控件,右击,选择"添加组"/"行组"/"父组",从 Tablix 组对话框中选择 Sno,重复以上操作,选择 Sname 为行的父组。选择[Grade]项并右击,选择"添加总计",得到如图 3-12 的设计界面。用中文修改表页眉中的文字,调整表格字体的位置和边框线条的粗线与颜色。

图 3-12 学生选课成绩单设计界面

(2) 报表窗体应用程序设计。启动 Visual Studio 2010 开发工具,选择"文件"/"新建"/"项目",从"已安装的模板"列表中选择"Reporting"/"报表应用程序",输入项目名称,点击"确定"。对弹出的报表向导对话窗口,选择"取消"。

在 Form1.cs 文件中添加引用 Microsoft.Reporting.WinForms。在 Form1_Load 函数中添加下面代码:

```
this.reportViewer1.ProcessingMode = ProcessingMode.Remote;
this.reportViewer1.ServerReport.ReportServerUrl =
            new Uri("http://localhost/reportserver");
this.reportViewer1.ServerReport.ReportPath = "/exampleDB 报表项目/Report1";
this.reportViewer1.RefreshReport();
```

按 F5 功能键,程序运行,显现图 3-10 窗口。

(3) 报表 Web 应用程序设计。在解决方案资源管理器中,右击项目名称,选择"添加"/"新建项"。从弹出的添加新建项窗口中,选择"Web"/"ASP.NET Web 程序",点击"确定"。

在 Default.aspx 的左下角选择"设计",从工具箱中,拖拽 ReportViewer 控件于设计窗口,再从工具箱的 AJAX Extensions 下拖拽 ScriptManager 控件到设计窗口。

打开 Default.aspx.cs 文件有,添加引用 Microsoft.Reporting.WebForms。在 Page_Load 函数中添加下列代码:

```
this.ReportViewer1.ProcessingMode = ProcessingMode.Remote;
this.ReportViewer1.ServerReport.ReportServerUrl =
```

```
            new Uri("http://localhost/reportserver");
this.ReportViewer1.ServerReport.ReportPath = "/exampleDB报表项目/Report1";
```

在解决方案资源管理器中,右击 Web 应用程序项目,选择"设为启动项",按 F5 功能键,浏览器启动并显示如图 3-11 所示页面。

【实验项目】

3.8 在本节实验示例的基础上,实现单科课程成绩单的打印。

实验项目 3.8

3.9 水晶报表应用程序设计

3.9.1 水晶报表基础

在 Visual Studio 2003、2005 和 2008 中,微软公司已将 Crystal 公司的报表插件集成到开发环境中。在 Visual Studio 2010 中,用户需要到 CrystalReport 官网下载"SAP Crystal Reports, version for Visual Studio 2010 — Standard"下载报表插件并安装。

借助 Crystal 报表模板,用户可以直接进行报表设计、预览和打印输出,也可以方便地将报表导出为 PDF 文档、Excel 文档、Word 文档等多种格式文档,并且在 Windows 应用程序或 Web 应用程序设计中使用水晶报表设计工具开发应用软件。

水晶报表分为标准报表、交叉报表和邮件标签三种类型报表,可以应用向导快速设计对应的报表。每个报表创建向导都由几个屏幕组成,通过这些屏幕指导创建指定的报表。许多报表创建向导有特定报表类型所独有的选项卡。

标准报表创建向导包括如何选择数据源和链接数据库表,帮助添加字段及指定要使用的分组、汇总(总计)及排序判据,以及创建图表和选择记录。用标准报表创建向导设计报表要经历以下屏幕:"数据"、"链接"、"字段"、"分组"、"汇总"、"组排序"、"图表"、"记录选择"和"报表样式"。

交叉表报表创建向导除标准向导包含的步骤外,还包括报表中的数据作为交叉表对象显示。其中的两个特殊屏幕("交叉表"和"网格样式")可帮助您创建交叉表本身并设置其格式。用交叉表报表创建向导设计报表要经历以下屏幕:"数据"、"链接"、"交叉表"、"图表"、"记录选择"、"网格样式"。

邮件标签报表创建向导允许创建设置为可打印在任意尺寸的邮件标签上的报表。可以使用"标签"屏幕选择一种商用标签类型,也可以自己定义用于任何多列样式报表的行列布局。用邮件标签报表创建向导设计报表要经历以下屏幕:"数据"、"链接"、"字段"、"标签"、

"记录选择"。

3.9.2 水晶报表设计器

在 Visual Studio 2008 专业版及更高版本中,系统中含有嵌入式水晶报表设计器。在环境中设计报表包括几个步骤:确定报表的总体结构;格式化并组织数据;使用数据进一步完善报表的组织,并显示希望包括的特定信息。

嵌入式水晶报表设计器如图 3-13 所示。窗体的左侧有工具箱,其中有 Text 对象、Line 对象和 Box 对象,这些对象提供了在报表中插入文本、直线和方框的能力;窗体的右侧为每段资源管理器和属性窗口,其中包含了数据库字段、公式、分组等报表资源,可以直接将对象拖拽到设计窗口;窗体的中间为水晶报表设计器,报表由多个部分组成,每一部分称为一个节(Section),分别为报表头、页眉、组头、详细资料、组尾、报表尾、页。

报表设计器提供了便捷的设计模式。鼠标点击节右侧的黑色三角,可折叠或展开节。鼠标右击节报表中对象均会弹出功能菜单。字段、直线和文本等对象均可随意拖动,属性窗口提供了对对象的精确控制。

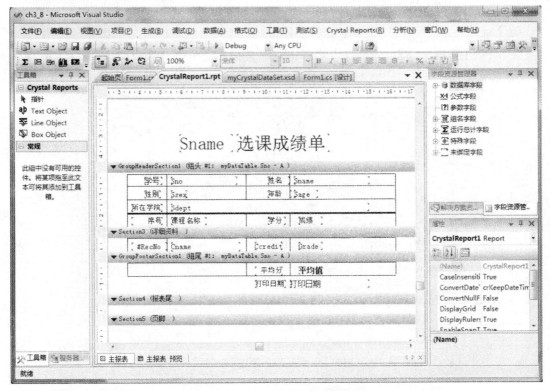

图 3-13 嵌入式水晶报表设计器

3.9.3 报表数据源

水晶报表与数据库数据的连接主要有直接连接和通过 ADO.NET 连接两种方法，前者称为拉模式，后者称为推模式。

在拉模式中，驱动程序将连接到数据库并根据需要将数据"拉"进来。使用这种模式时，与数据库的连接和为了获取数据而执行的 SQL 命令都同时由水晶报表本身处理，不需要开发人员编写代码。如果在运行时无须编写任何特殊代码，则使用拉模式。

在推模式中，与拉模式相反，开发人员需要编写代码以连接到数据库，执行 SQL 命令以创建与报表中的字段匹配的记录集或数据集，并且将该对象传递给报表。该方法可以将连接共享置入应用程序中，并在水晶报表收到数据之前先将数据筛选出来。

推模式由于使用 ADO.NET 数据集作为报表数据源，在数据传给报表前可以利用 ADO.NET 技术对数据进行处理，具有很强的灵活性。但在报表预览时无法直接浏览实际数据，系统用一些随机数据填充报表，只有在程序运行时，才能显示实际数据。

3.9.4 水晶报表示例程序

【实验 3-8】 分页打印 exampleDB 数据库中学生选课成绩单，成绩单中有学生个人信息、所选课程、每门课的学分、成绩和平均分，如图 3-14 所示。

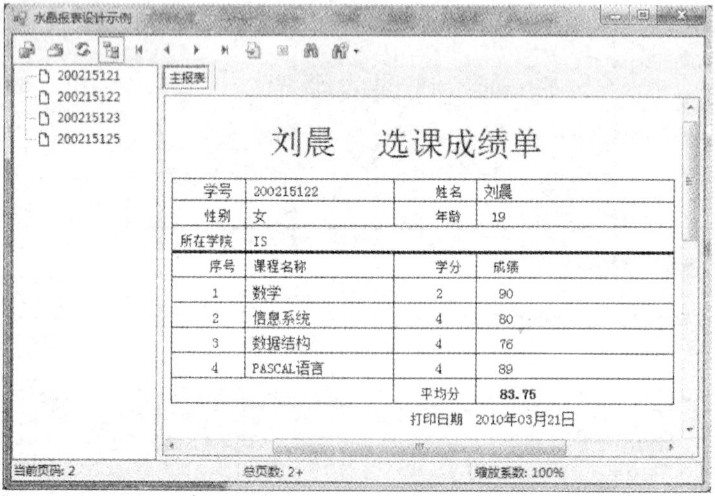

图 3-14 水晶报表设计示例

实验步骤：

（1）创建 Windows 应用程序项目。在服务器资源管理器中，创建新的数据连接，连接 exampleDB 数据库，名称为 exampleDBConnectionString。

（2）建立数据集。按组合键"Ctrl+Shift+A"，弹出"添加"对话框，选择"数据集"，名称设为 myCrystalDataSet.xsd，点击"添加"按钮。在数据集设计器窗口中点击鼠标右键，从弹出菜单中选择"TableAdapter"，弹出如图 3-15 所示的配置向导。

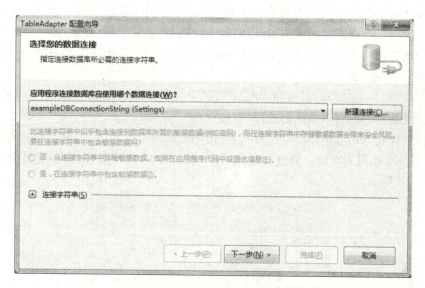

图 3-15　TableAdapter 配置向导

在"选择您的数据连接"页上，从当前可用连接的列表中选择数据连接，或选择"新建连接"创建一个新连接。由于应用程序已经建立了数据库连接，默认的连接选项即为 exampleDBConnectionString 连接项，点击"下一步"。

在"选择命令类型"页上，保持默认选项"使用 SQL 语句"，点击"下一步"。

在"输入 SQL 语句"页上，点击"查询生成器"按钮，添加 Student、SC、Coruse 三个表，生成如下 SQL 语句：

```
SELECT    STUDENT.SNO, STUDENT.SNAME, STUDENT.SSEX, STUDENT.SDEPT,
STUDENT.SAGE, SC.GRADE, COURSE.CNAME, COURSE.CCREDIT
FROM   STUDENT INNER JOIN
         SC ON STUDENT.SNO = SC.SNO INNER JOIN
         COURSE ON SC.CNO = COURSE.CNO
```

在"选择要生成的方法"页上，不改变"填充 DataTable"和"返回 DataTable"设置，点击"下一步"。

在"向导结果"页上,点击"完成"按钮,结束数据集的创建。

在 myCrystalDataSet 数据集设置器中,修改 DataTable1 为 myDataTable,DataTable1DataAdapter 为 myDataTableDataAdapter。

(3) 创建水晶报表文档。按组合键"Ctrl+Shift+A",弹出"Crystal Reports 库"对话框,使用缺省选项"使用报表向导"和"标准"表格,点击"确定"按钮。

在"数据"页上,从"可用数据源"中选择"项目数据",再展开"ADO.NET 数据集",选中"myDataTable",点击">"按钮,将表选入到"选定的表"中,再点击"下一步"。

在"字段"页上,点击">>"按钮,将所有字段选至"要显示的字段"中。点击"下一步"。

在"分组"页上,从"可用字段"中选中 myDataTable.Sno,点击">"按钮,将该字段选入"分组依据"。点击"下一步"。

在"汇总"页上,点击"<<"按钮,"汇总字段"中只剩 myDataTable.Sno,点击"下一步"。

在"记录选择"页上,不做任何选择,直接点击"下一步"。

在"报表样式"页上,选择"标准"样式,点击"完成"。

(4) 修改水晶报表文档。通过可视化操作,添加线条,修改文字,调整字段位置,设计表格如图 3-16 所示。

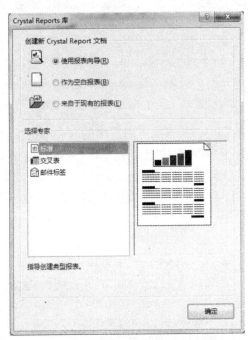

图 3-16 创建水晶报表对话框

(5) 增加课程序号。在"字段资源管理器"中,右击"运行总计字段",选择"新建"。从

第 3 章　数据库应用程序设计基础

"可用表和字段"中，选中 myDataTable.Cname，点击">"按钮，汇总字段为 myDataTable.Cname，汇总类型选择为"计数"，运行总计名称改为 RecNo。"求值"项设为"对每个记录"，"重置"项设为"组更改时"。点击"确定"，拖拽 RecNo 到相应位置。

（6）设置分页。在报表设计窗口，右击"组头"栏，选择"节专家"。在"公用"页上，"保持在一起"复选框为选中状态，在"分页"页上，选择"之前新建页"复选框为选中状态，其余为未选中状态。

（7）平均值的插入。右击"组尾"右边表格空白处，选择"插入"/"汇总". 在"选择要汇总的字段"项下选择"myDataTable.Grade"，"计算此汇总"选择"平均值"，"汇总位置"选择"组 #1:myDataTable.Sno－A"，点击"确定"。

（8）从工具箱中拖拽 CrystalReportViewer 控件至 Form1 窗体中，双击窗体，添加连接数据库、创建数据集和水晶报表对象的代码。

程序代码：

```
public partial class Form1 : Form
{
    SqlConnection myConnection;
    myCrystalDataSet myDataSet;
    SqlDataAdapter myDAdapter;
    public Form1()
    {
        InitializeComponent();
    }
    private void Form1_Load(object sender, EventArgs e)
    {
        var myConnString = Properties.Settings.Default.exampleDBConnectionString;
        //连接数据库,填充数据表
        myConnection = new SqlConnection(myConnString);
        myDataSet = new myCrystalDataSet();
        myDAdapter = new SqlDataAdapter("SELECT    Student.Sno, Student.Sname,
            Student.Ssex,Student.Sdept,Student.Sage, SC.Grade, Course.Cname,
            Course.Ccredit, Course.CcoverPhoto FROM Student INNER JOIN  SC ON
            Student.Sno = SC.Sno INNER JOIN Course ON SC.Cno = Course.Cno",
            myConnection);
        myDataSet.EnforceConstraints = false;
        myDAdapter.Fill(myDataSet, "myDataTable");
```

```
            //创建报表对象,设置数据源和报表源
        CrystalReport1 myCR= new CrystalReport1();
        myCR.SetDataSource(myDataSet);
        crystalReportViewer1.ReportSource = myCR;
    }
}
```

【实验项目】
3.9 在本节实验示例的基础上,实现单科课程成绩单的打印。

实验项目 3.9

3.10 安装程序制作

3.10.1 Microsoft Windows Installer 程序简介

Windows Installer 是作为 Windows 操作系统的组成部分随带的安装和配置服务。它基于数据驱动模型,在一个软件包中提供所有安装数据和指令。传统的脚本安装程序基于过程模型,为应用程序安装提供脚本指令。脚本安装程序强调"如何"安装,而 Windows Installer 则强调安装"什么"。

Windows Installer 为每台计算机都保留一个信息数据库,其中的信息与它所安装的每个应用程序有关,包括文件、注册表项和组件。卸载应用程序时,将检查数据库以确保在移除该应用程序前没有其他应用程序依赖于文件、注册表项或组件。这样可防止在移除一个应用程序后中断另一个应用程序。Windows Installer 还支持自我修复,即应用程序能够自动重新安装因用户误删除而丢失的文件。此外,Windows Installer 具有回滚的功能。例如,如果应用程序依赖于某个特定的数据库,而在安装过程未找到该数据库,则可以中止安装,计算机系统则返回到安装前的状态。

Visual Studio 2010 中的安装部署工具建立在 Windows Installer 的基础之上,为应用程序的部署和维护提供了丰富的功能。

3.10.2 创建安装程序

在 Visual Studio 2010 中,可通过建立安装项目来创建安装程序,以便分发应用程序。最终的 Windows Installer(.msi)文件包含应用程序、任何依赖文件以及有关应用程序的信息(如注册表项和安装说明)。

在 Visual Studio 中,有两种类型的安装项目,即"安装"项目和"Web 安装"项目。"安装"项目与"Web 安装"项目之间的区别在于安装程序的部署位置:"安装"项目将文件安装到目标计算机的文件系统中;而"Web 安装"项目将文件安装到 Web 服务器的虚拟目录中。此外,还提供了"安装向导"以简化创建"安装"项目或"Web 安装"项目的过程。

3.10.3 创建卸载程序

利用安装项目创建的安装程序中不含卸载程序,程序卸载需要打开系统"控制面板"中的"卸载程序"功能,找到所安装的程序,双击执行卸载。

程序卸载是调用 Windows Installer 程序并传递相应参数完成。该程序的文件名为 msiexec.exe,保存在系统盘下,路径为"系统盘符:\Windows\System32"。在命令提示符窗口执行 msiexec.exe /? 可以获得 Windows Installer 程序的使用帮助。

在实验 3-10 中,介绍了一种卸载程序的方法。其主要思想是在安装项目中添加一个具有调用 msiexec.exe 卸载指定程序功能的控制台应用程序。

3.10.4 安装程序的设计示例程序

【实验 3-10】 在实验 3-3 项目中添加安装程序。

实验步骤:

(1) 打开实验 3-3 项目。在解决方案资源管理器中,右击"解决方案",选择"添加"/"新建项目",弹出"添加新建项目"对话框。在"已安装的模板"列表中;打开"其他项目类型",选择"安装和部署"/"Visual Studio Installer",再选择"安装项目",设项目名称为"Setup1",点击"确定"按钮。

(2) 设计卸载程序。在解决方案资源管理器中,右击"解决方案",从弹出菜单中选择"添加"/"新建项",选择添加"控制台应用程序",命名为 Uninstall。在 Program.cs 中,添加程序代码,详见程序代码部分。程序中的{A3B32F43-1779-459C-A791-76711597B571}为 ProductCode,点击"ch3_10"项目,从属性窗口即可复制到 ProductCode 的值。

右击"Uninstall"项目,从弹出菜单中选择"生成",在 Uninstall 文件夹的 bin\Debug 子文件夹生成可执行文件 Uninstall.exe。

(3) 在"文件系统(Setup1)"页中,右击"应用程序文件夹",从弹出菜单中选择"添加"/"文件",弹出"添加文件"对话框。从原项目生成的 Debug 文件夹下,选择 ch3_3.exe 和 ch3_3.exe.config 文件添加到"应用程序文件夹"。再从 Uninstall 项目下,选择"Uninstall.exe"文件,添加到"应用程序文件夹"。如图 3-17 所示。

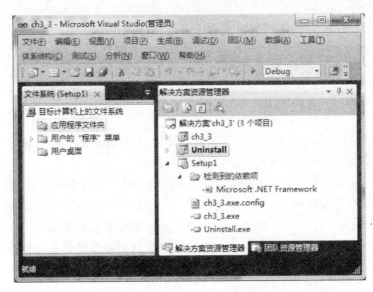

图 3-17 安装制作示例程序设计界面

(4) 在"文件系统(Setup1)"中,右击"用户的"程序"菜单,从弹出菜单中选择"添加"/"文件夹",输入文件夹名为"ch3_3"。点击该文件夹,在右边窗格中右击,从弹出菜单中选择"创建新的快捷方式",在"选择项目中的项"对话框中选择"ch3_3.exe"文件,创建 ch3_3.exe 的快捷方式。同样为"Uninstall.exe"文件创建 Uninstall 快捷方式。

(5) 点击"用户桌面",为"ch3_3.exe"文件在桌面创建快捷方式。

(6) 在解决方案资源管理器中,点击"Setup1"项目,在属性窗口中为下列项修改属性值或输入相应参数:作者(Author)、所有用户(InstallAllUsers)、制造商(Manufacturer)、产品名称(ProductName)、标题(Title)。

(7) 在解决方案资源管理器中,右击"Setup1"项目,从弹出菜单中选择"属性",出现"ch3_10 属性页"对话框。在对话框中,设输出文件名为"Debug\Setup.msi",点击"系统必备"按钮可选择安装系统所必备的组件和安装位置。此外,右击"Setup1"项目,在弹出菜单中还有"生成"、"视图"、"安装"和"卸载"功能,"视图"项中含有"用户界面"项,可对安装程序的交互界面进行设计。

(8) 使用"生成"功能,完成项目的生成。使用"安装"功能,测试设计是否正确。使用"卸载"功能,删除所安装的软件。最后,"Setup1"文件夹的"Debug"子文件夹中找到制作成功的安装文件。

程序代码:

```
namespace Uninstall
```

```
    {
        class Program
        {
            static void Main(string[] args)
            {
                string path = System.Environment.SystemDirectory + "\\msiexec.exe";
                if(System.IO.File.Exists(path))
                {
                    System.Diagnostics.Process.Start(path, "/x {5F2CFA79-F319-4055-BC94-D455A6FEB5F5} /passive");
                }
            }
        }
    }
```

【实验项目】

3.10 为实验 3-6(见 3.6.2 节)设计安装与卸载程序并测试。

第 4 章 学生成绩管理系统

数据库应用软件设计通常要经历需求分析、概念结构设计、数据库设计以及应用系统开发等过程,其中数据库设计在系统开发中占有极其重要的地位。

本章以一个简化了的学生成绩管理系统设计为例,介绍数据库应用程序设计的主要步骤和 WinForm 桌面应用软件的实现技术。系统后台数据库采用 SQL Server 2008,前台采用 Visual C♯ 2010 为开发工具,软件具有数据录入、删改、查询、报表打印等基本功能。

4.1 系统设计

4.1.1 系统需求及功能概述

学生成绩管理系统是将学生的各科成绩利用计算机进行加工处理,以实现课程成绩的电子化,提高工作效率。

学生成绩管理系统强化了学生成绩管理的职能,涵盖了学生管理、成绩管理、课程管理、教师管理等主要功能,特别是对学生成绩管理工作进行了概括和提炼,使成绩管理系统管理工作日益规范化和科学化。

成绩管理系统实现了查询功能,可以在全校范围内查找学生、课程、教师和成绩等信息;实现了添加信息功能,可以添加学生信息、课程信息、教师信息和录入成绩等;实现了删除功能,可以删除学生、课程、教师和成绩等信息;还实现了成绩统计、密码修改等功能。

成绩管理系统用户分为管理员、教师和学生。管理员的权限为修改本人密码,浏览、添加、删除、修改学生信息、教师信息、课程信息和成绩信息。教师的权限为修改本人密码,查看、修改和添加自己的课程成绩。学生的权限为修改本人密码,浏览自己所学课程的成绩。系统功能模块划分如图 4-1 所示。

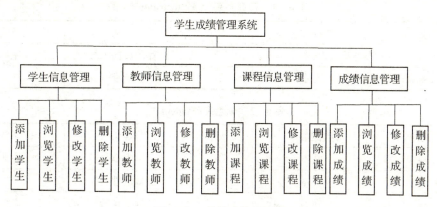

图 4-1 系统功能结构图

4.1.2 概念结构设计

学生成绩管理系统所要处理的数据有学生个人信息、教师信息、课程信息以及考试成绩。分析它们之间的关系,得到系统的E-R图(如图4-2所示),其中每个实体及属性如下:

学生:学号、姓名、性别、年龄、系科、登录密码

课程:课程号、课程名、先修课程号、学分

教师:工号、姓名、登录密码

管理员:登录名、密码

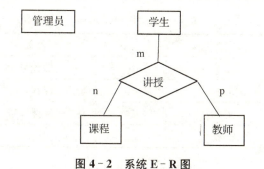

图 4-2 系统 E-R 图

4.1.3 数据库设计

系统数据库名为:StudentManger,数据库中包括:管理员表(Administrator)、学生表(Student)、教师表(Teacher)、课程表(Course)、成绩表(Grade)。

各个表的数据结构如表 4-1~4-5 所示。

表 4-1 管理员表(Administrator)

字段名称	数据类型	大小	含义	备注
userid	文本	10	管理员标识号	主码
userpassword	文本	10	登录密码	

表 4-2 学生表(Student)

字段名称	数据类型	大小	含义	备注
Sno	文本	10	学生学号	主码
Sname	文本	10	学生姓名	
Sage	整数		学生年龄	
SSex	文本	2	学生性别	
Sdept	文本	20	学生系科	
Spassword	文本	12	学生密码	

表 4-3 教师表(Teacher)

字段名称	数据类型	大小	含义	备注
Tno	文本	10	教师工号	主码
Tname	文本	10	教师姓名	
Tpassword	文本	12	教师密码	

表 4-4 课程表(Course)

字段名称	数据类型	大小	含义	备注
Cno	文本	10	课程号	主码
Cname	文本	20	课程名	
Cpno	文本	10	先修课程号	
Ccredit	整数		学分	

表 4-5 成绩表(Grade)

字段名称	数据类型	大小	含义	备注
Sno	文本	10	学号	主码
Cno	文本	10	课程号	
Tno	文本	10	教师工号	
Grade	整数	20	成绩	

表与表之间的关系：

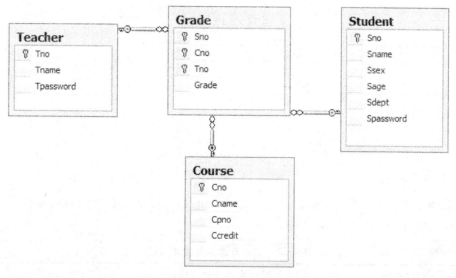

图 4-3　表与表关系图

Grade 表中外码说明如表 4-6 所示。

表 4-6　外码说明

外码	参照关系	更新规则	删除规则
Sno	参照 Student(Sno)	级联	级联
Cno	参照 Course(Cno)	级联	级联
Tno	参照 Teacher(Tno)	级联	级联

4.2　系统详细设计

启动 Visual Studio 2008，选择"文件"|"新建"|"项目"命令，新建学生成绩管理应用程序。下面介绍各窗体的程序设计。

4.2.1　登录界面设计

登录界面的作用就是判断用户身份的合法性。输入用户名和密码，选择用户身份，若用户名、密码正确并且和身份相符，单击[确定]按钮，即可进入不同身份对应的主界面，否则不能进入。登录界面效果，如图 4-4 所示。

图 4-4 登录界面

该窗体中设计了 2 个 TextBox 控件、1 个 ComboBox 控件和 2 个 Button 控件。各个控件的名称、作用如表 4-7 所示。

表 4-7 登录界面主要控件设计

控件类型	控件名称	作用
TextBox	usernamebox	用户名
	upwdbox	用户密码
comboBox	Identitybox	用户身份
button	okbtn	确定
	cancelbtn	取消

登录界面部分代码：

1. 该代码中定义了公共变量 userid 用来向其他窗体传递数据。当单击"确定"按钮时，触发 okbtn_Click 事件。

```
public static string userid;
private void okbtn_Click(object sender, EventArgs e)
{
    SqlConnection cn = new SqlConnection(" data source =. ; initial catalog = Studentmanger; integrated security=true");
    SqlCommand cmd = new SqlCommand();
    SqlDataReader datareader;
    cmd.Connection = cn;
    if (cn.State == ConnectionState.Closed)
        cn.Open();
```

```csharp
            string str;
            if (identybox.SelectedIndex == 2)
            {   str=string.Format("select * from administrator where userid='{0}' and userpassword='{1}'", usernamebox.Text.Trim(), upwdbox.Text.Trim());
                cmd.CommandText = str;
                userid = usernamebox.Text;
                datareader = cmd.ExecuteReader();
                if (datareader.Read())
                {
                    main main = new main();
                    this.Hide();
                    main.Show();
                }
                else
                {
                    MessageBox.Show("请正确输入!");
                }
            }
            if (identybox.SelectedIndex == 1)
            {
                str = string.Format("select * from student where sno='{0}' and spassword='{1}'", usernamebox.Text.Trim(), upwdbox.Text.Trim());
                cmd.CommandText = str;
                userid = usernamebox.Text.Trim();
            datareader = cmd.ExecuteReader();
                if (datareader.Read())
                {
                    studentmain studentmain = new studentmain();
                    this.Hide();
                    studentmain.Show();
                }
                else
                {
                    MessageBox.Show("请正确输入用户名,密码!");
                }
            }
            if (identybox.SelectedIndex == 0)
```

```
            {
                str = string.Format("select * from teacher where tno='{0}' and tpassword='{1}'",
usernamebox.Text.Trim(), upwdbox.Text.Trim());
                cmd.CommandText = str;
                userid = usernamebox.Text.Trim();
                datareader = cmd.ExecuteReader();
                if (datareader.Read())
                {
                    teachermain f1 = new teachermain();
                    this.Hide();
                    f1.Show();
                }
                else
                {
                    MessageBox.Show("请正确输入用户名,密码!");
                }
            }
        }
```

2. 登录时,学生使用学号登录,教师使用工号登录,管理员使用用户名登录,登录成功后分别转向对应的学生主窗口、教师主窗口、管理员主窗口。

当单击"取消"按钮时,触发 cancelbtn_Click 事件,关闭当前窗体。代码如下:

```
private void cancelbtn_Click(object sender, EventArgs e)
{
    this.Close();
}
```

4.2.2 管理员主界面设计

选择管理员身份登录后进入管理员主界面,如下图 4-5 所示。在该界面中设计了一个 menuStrip 控件,并且设置窗体的 IsMdiContainer 属性为 True。

图 4-5 管理员主界面

该主界面的作用就是显示管理员所拥有的所有的功能菜单项,当用户单击相应的菜单项时,打开对应的模块窗口。

管理员主界面 Main 部分代码

1. 在管理员主界面中选择"系统管理\添加管理员"命令菜单,就会触发添加管理员 ToolStripMenuItem_Click 事件,进入添加管理员界面。代码如下:

```csharp
private void 添加管理员ToolStripMenuItem_Click(object sender, EventArgs e)
{
    addadmin f1 = new addadmin();
    for (int x = 0; x < this.MdiChildren.Length; x++)
    {
        Form tempChild = (Form)this.MdiChildren[x];
        tempChild.Close();
    }
    f1.WindowState = FormWindowState.Normal;
    f1.MdiParent = this;
    f1.Show();
}
```

2. 在管理员主界面中选择"重新登录"命令菜单,就会触发添加管理员 ToolStripMenuItem_Click 事件,进入登录界面。代码如下:

```csharp
private void 重新登录ToolStripMenuItem_Click(object sender, EventArgs e)
{
    Login f1 = new Login();
    this.Close();
    f1.Show();
}
```

3. 在管理员主界面中选择"退出"命令菜单,就会触发退出 ToolStripMenuItem_Click 事件,整个程序退出。代码如下:

```csharp
private void 退出ToolStripMenuItem_Click(object sender, EventArgs e)
{
    Application.Exit();
}
```

点击其他菜单命令分别转入各自对应的窗口,触发事件代码与"添加管理员"菜单触发事件代码基本相同,这里不在赘述。

4.2.3 系统管理

4.2.3.1 添加管理员

在管理员主界面中选择"系统管理\添加管理员"命令菜单,即可进入添加管理员界面,如图 4-6 所示。该窗体中需要提供的信息包括:管理员用户名、密码、确认密码。信息录入后单击"确定"按钮,如果新用户名不存在并且密码和确认密码相同则向数据库中插入一条信息。

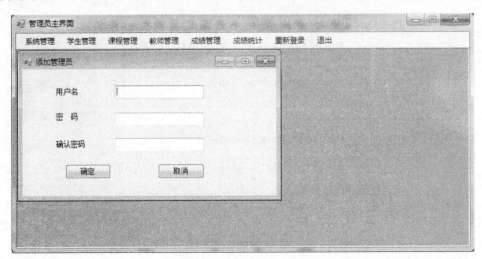

图 4-6 添加管理员界面

在该窗体中主要设计了 3 个 TextBox 控件和 2 个 Button 控件。各个控件的名称、作用如表 4-8 所示。

表 4-8 添加管理员界面控件设计

控件类型	控件名称	作用
TextBox	usernamebox	用户名
	passwordbox	用户密码
	password2box	确认密码
Button	okbtn	确定
	cancelbtn	取消

添加管理员界面代码:

1. 单击"确定"按钮时触发 okbtn_Click 事件,第一步判断信息是否完整;若完整,第二步判断两次密码是否相同;第三步判读用户名是否存在,若否,则插入该信息。代码如下:

```csharp
private void okbtn_Click(object sender, EventArgs e)
{
    if (this.usernamebox.Text.Trim() == "" || this.passwordbox.Text.Trim() == "" || this.password2box.Text.Trim() == "" )
    {
        MessageBox.Show("请输入完整信息!","警告");
    }
    else
    {
        if (this.passwordbox.Text.Trim() != this.password2box.Text.Trim())
        {
            MessageBox.Show("两次密码输入不一致!","警告");
        }
        else
        {
            SqlConnection cn = new SqlConnection("server=.;database=studentmanger;integrated security=true;");
            cn.Open();
            string sql = "select * from administrator where userID = '" + this.usernamebox.Text.Trim() + "'";
            SqlCommand cmd = new SqlCommand(sql, cn);
            if (null == cmd.ExecuteScalar())
            {
                string sql1="insert into administrator (userid,userpassword)" + "values ('" + this.usernamebox.Text.Trim() + "','" + this.passwordbox.Text.Trim() + "')";
                cmd = new SqlCommand(sql1, cn);
                cmd.ExecuteNonQuery();
                MessageBox.Show("添加用户成功!","提示");
                this.Close();
            }
            else
                MessageBox.Show("用户名" + this.usernamebox.Text.Trim() + "已经存在!","提示");
            cn.Close();
        }
    }
}
```

2. 单击"取消"按钮时,将触发 cancel_Click 事件,关闭当前窗体。代码如下:

```
private void cancelbtn_Click(object sender, EventArgs e)
{
    this.Close();
}
```

4.2.3.2 修改密码

在管理员主界面中选择"系统管理\修改密码"命令菜单,即可进入修改密码界面,如图 4-7 所示。每个管理员只能修改自己的密码,所以需要利用 Login 窗体中传递的公共变量 userid,将该值传递给 usernamebox 的 Text 属性,并且设置 usernamebox 的 ReadOnly 属性为 True。

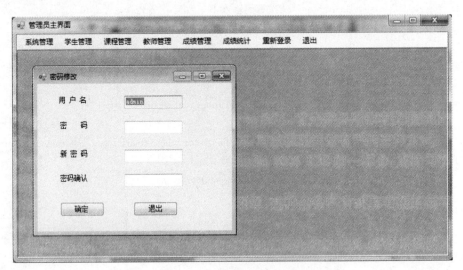

图 4-7 修改管理员密码界面

在该窗体中主要设计了 4 个 TextBox 控件和 2 个 Button 控件。各个控件的名称、作用如表 4-9 所示。

表 4-9 修改密码控件设计

控件类型	控件名称	作用
TextBox	usernamebox	用户名(注:ReadOnly 属性为 True)
	psswdbox	用户密码
	newpsdbox	新密码
	new2psdbox	确认新密码

(续表)

控件类型	控件名称	作用
Button	okbtn	确定
	exitbtn	退出

修改密码界面代码:

1. 窗体加载时触发 modifypassword_Load 事件。传递 Login 窗体中的变量值 userid。代码如下:

```csharp
private void modifypassword_Load(object sender, EventArgs e)
{
    usernamebox.Text = Login.userid;
}
```

2. 单击确定按钮时触发 okbtn_Click 事件。代码如下:

```csharp
private void okbtn_Click(object sender, EventArgs e)
{
    if ((usernamebox.Text.Trim() == "")||(psswdbox.Text.Trim() == "")||(newpsdbox.Text.Trim() == "") ||( new2psdbox.Text == ""))
        MessageBox.Show("请填写完整信息","提示", MessageBoxButtons.OKCancel, MessageBoxIcon.Warning);
    else
    {
        cn.Open();
        SqlCommand cmd=new SqlCommand();
        string str = string.Format("select * from administrator where userid='{0}' and userpassword='{1}' ",usernamebox.Text.Trim(),psswdbox.Text.Trim());
        cmd.CommandText=str;
        cmd.Connection=cn;
        if(null!=cmd.ExecuteScalar())
        {
            if(newpsdbox.Text.Trim()!=new2psdbox.Text.Trim())
                MessageBox.Show("两次密码不一致","提示");
            else
            {
                string str2=string.Format("update administrator set userpassword='{0}' where userid='{1}'",newpsdbox.Text.Trim(),usernamebox.Text.Trim());
```

```
                cmd.CommandText=str2;
                cmd.ExecuteNonQuery();
                MessageBox.Show("密码修改成功");
                this.Close();
            }
        }
        else
        {
            MessageBox.Show("密码错误!","提示");
        }
        cn.Close();
    }
}
```

4.2.4 学生管理

4.2.4.1 学生查询、删除

在管理员主界面中选择"学生管理\学生查询删除"命令菜单,即可进入学生查询界面,如图 4-8 所示。

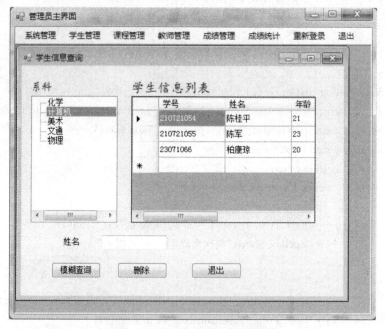

图 4-8 学生信息查询删除界面

该窗体中主要设计了 1 个 TextBox 控件、3 个 Button 控件、1 个 DataGridView 控件和 1 个 TreeView 控件。各个控件的名称、作用如表 4-10 所示。

表 4-10　学生查询界面控件设计

控件类型	控件名称	作用
TextBox	snameBox	学生姓名
Button	selbtn	模糊查询
	delbtn	删除
	exitbtn	退出
DataGridView	dataGridView1	显示学生信息
TreeView	treeView1	显示系科列表

学生查询删除界面主要代码：

1. 窗体加载时触发 studentinfo_Load 事件，通过该事件向 TreeView 控件中增加专业结点使专业以树的形式展现。代码如下：

```
private void studentinfo_Load(object sender, EventArgs e)
{
    string str = "select distinct sdept from student ";
    if (cn.State == ConnectionState.Closed)
        cn.Open();
    SqlCommand cmd = new SqlCommand(str, cn);
    SqlDataReader dr = cmd.ExecuteReader();
    while (dr.Read())
    {
        TreeNode node = new TreeNode();
        node.Text = dr.GetString(0).ToString();
        treeView1.Nodes.Add(node);
    }
    dr.Close();
    cn.Close();
}
```

2. 点击 TreeNode 中的专业后触发 treeView1_AfterSelect 事件，在 dataGridView1 中显示所点专业的所有学生信息。代码如下：

```
private void treeView1_AfterSelect(object sender, TreeViewEventArgs e)
```

```csharp
{
    string sql = "select Sno 学号,Sname 姓名,Sage 年龄,Ssex 性别,sdept 系科 from student where sdept='" + e.Node.Text.ToString() + "'";
    myadapter = new SqlDataAdapter(sql,cn);
    ds = new DataSet();
    ds.Clear();
    if (cn.State == ConnectionState.Closed)
        cn.Open();
    myadapter.Fill(ds, "student");
    dataGridView1.DataSource = ds.Tables[0].DefaultView;
    cn.Close();
}
```

3. 在 snamebox 中输入姓名或姓名中的某一个字符后点击模糊查询触发 selbtn_Click 事件,通过在 select 语句中使用"%"通配符实现模糊查询。代码如下:

```csharp
private void selbtn_Click(object sender, EventArgs e)
{
    string s = ds.Tables[0].Rows[n][4].ToString();
    string str = string.Format("select sno 学号,sname 姓名,sage 年龄,sdept 系科 from student where sname like '%{0}%' and sdept='{1}'",snamebox.Text.Trim(),s);
    if(cn.State==ConnectionState.Closed)
    cn.Open();
    SqlCommand cmd = new SqlCommand(str, cn);
    ds = new DataSet();
    ds.Clear();
    myadapter = new SqlDataAdapter(cmd);
    myadapter.Fill(ds, "student");
    dataGridView1.DataSource = ds.Tables[0];
}
```

4. 单击 dataGridView1 中的某个元组后,触发 dataGridView1_CellClick 事件,用外部变量 n 记录下光标所指向的行。代码如下:

```csharp
private void dataGridView1_CellClick(object sender, DataGridViewCellEventArgs e)
{
    n = e.RowIndex;
}
```

5. 单击删除按钮触发 delbtn_Click 事件,将删除 dataGridView1 中被选中的第 n 行。

代码如下:

```
private void delbtn_Click(object sender, EventArgs e)
{
    SqlCommandBuilder cmdbuilder = new SqlCommandBuilder(myadapter);
    if(MessageBox.Show("真的删除？删除后无法恢复","提示",MessageBoxButtons.YesNo,MessageBoxIcon.Warning) == DialogResult.Yes)
    {
        DataRow delrow = ds.Tables[0].Rows[n];
        delrow.Delete();
        myadapter.Update(ds,"student");
        cn.Close();
        MessageBox.Show("删除成功");
    }
}
```

4.2.4.2 添加学生信息

在管理员主界面中选择"学生管理\添加学生"命令菜单,即可进入添加学生界面,如图4-9所示。

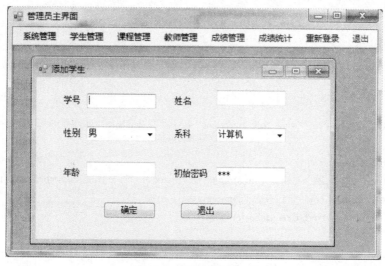

图4-9 添加学生界面

该窗体中主要有4个TextBox控件、2个Button控件、2个ComboBox控件。各个控件的名称、作用如表4-11所示。

表 4-11 添加学生界面控件设计

控件类型	控件名称	作用
TextBox	Snobox	学生学号
	snamebox	学生姓名
	sagebox	学生年龄
	spsdbox	初始密码(text 属性设为 111)
Button	okbtn	确定
	cancelbtn	退出
ComboBox	ssexbox	学生性别
	sdeptbox	学生系科

添加学生界面主要代码：

单击确定按钮，触发 okbtn_Click 事件，若已经输入完整信息，并且数据库中没有相同的学号，则完成向数据库中插入一行学生信息。代码如下：

```
private void okbtn_Click(object sender, EventArgs e)
{
    if (this.snobox.Text.Trim() == "" || this.snamebox.Text.Trim() == "" || this.sagebox.Text.Trim() == "")
    {
        MessageBox.Show("请输入完整信息!","警告");
    }
    else
    {
        SqlConnection cn = new SqlConnection("server=. ;database=studentmanger;integrated security=true;");
        cn.Open();
        string sql = "select sno from student where sno = '" + this.snobox.Text.Trim() + "'";
        SqlCommand cmd = new SqlCommand(sql, cn);
        if (null == cmd.ExecuteScalar())
        {
            string sql1 = "insert into student(sno,sname,ssex,sage,sdept,spassword)" + "values ('" +
this.snobox.Text.Trim() + "','" +this.snamebox.Text.Trim() + "','"+
this.ssexbox.Text.Trim() + "','"+this.sagebox.Text.Trim()+",'" +this.sdeptbox.Text.Trim()
+ "','"+ this.spsdbox.Text.Trim()+"')";
```

```
                cmd = new SqlCommand(sql1, cn);
                cmd.ExecuteNonQuery();
                MessageBox.Show("添加用户成功!","提示");
                this.Close();
            }
            else
                MessageBox.Show("用户名" + this.snobox.Text.Trim() + "已经存在!","提示");
                cn.Close();
        }
```

4.2.4.3 学生信息修改

在管理员主界面中选择"学生管理\学生信息修改"命令菜单,即可进入学生信息修改界面,如图 4-10 所示。

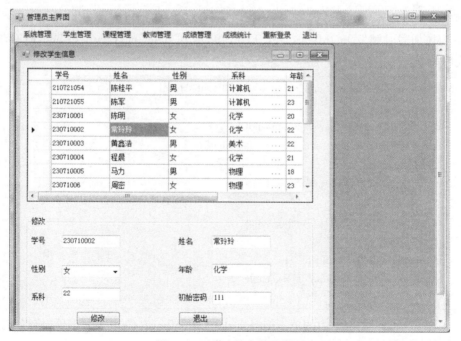

图 4-10 学生信息修改界面

该窗体中主要了设计了 5 个 TextBox 控件、2 个 Button 控件、1 个 ComboBox 控件和 1 个 DataGridView 控件。各个控件的名称、作用如表 4-12 所示。

表4-12 学生查询界面控件设计

控件类型	控件名称	作用
TextBox	snobox	学生学号
	snamebox	学生姓名
	sagebox	学生年龄
	sdeptbox	学生系科
	psswdbox	学生密码
Button	alterbtn	确定
	exitbtn	退出
ComboBox	ssexbox	学生性别
DataGridView	dataGridView1	显示学生信息

学生信息修改界面主要代码：

1. 窗体加载时触发 modifystudent_Load 事件，在 dataGridView1 种显示所有学生信息。代码如下：

```
private void modifystudent_Load(object sender, EventArgs e)
{
    SqlCommand cmd = new SqlCommand();
    cmd.CommandText = "select * from student";
    cmd.Connection = cn;
    if (cn.State == ConnectionState.Closed)
        cn.Open();
    myadapter = new SqlDataAdapter(cmd);
    ds.Clear();
    myadapter.Fill(ds, "student");
    dataGridView1.DataSource = ds.Tables[0];
}
```

2. 单击 dataGridView1 中的单元格触发 dataGridView1_CellClick 事件，将当前选择的学生信息在下面对应的文本框或组合框显示。代码如下：

```
private void dataGridView1_CellClick(object sender, DataGridViewCellEventArgs e)
{
    n = e.RowIndex;
    snobox.Text = dataGridView1.Rows[e.RowIndex].Cells[0].Value.ToString();
```

```
        snamebox.Text = dataGridView1.Rows[e.RowIndex].Cells[1].Value.ToString();
        ssexbox.Text = dataGridView1.Rows[e.RowIndex].Cells[2].Value.ToString();
        sagebox.Text = dataGridView1.Rows[e.RowIndex].Cells[3].Value.ToString();
        sdeptbox.Text = dataGridView1.Rows[e.RowIndex].Cells[4].Value.ToString();
        psswdbox.Text = dataGridView1.Rows[e.RowIndex].Cells[5].Value.ToString();
    }
```

3. 在文本框或组合框输入或选择新值后点击修改触发 alterbtn_Click 事件,更新数据库。代码如下:

```
    private void alterbtn_Click(object sender, EventArgs e)
    {
        SqlCommandBuilder comdbuilder = new SqlCommandBuilder(myadapter);
        DataTable mytable = ds.Tables[0];
        DataRow updaterow;
        updaterow=mytable.Rows[n];
        updaterow.BeginEdit();
        updaterow["sno"] = snobox.Text;
        updaterow["sname"] = snamebox.Text;
        updaterow["ssex"] = ssexbox.Text;
        updaterow["sage"] = sagebox.Text;
        updaterow["sdept"] = sdeptbox.Text;
        updaterow["spassword"] = psswdbox.Text;
        updaterow.EndEdit();
        myadapter.Update(ds, "student");
        ds.AcceptChanges();
        MessageBox.Show("更新成功!");
    }
```

4.2.5 课程管理、教师管理、成绩管理

课程管理、教师管理、成绩管理与学生管理相似,具体代码参见随书光盘,这里不再赘述。

4.2.6 成绩统计

在管理员主界面中选择"成绩统计"命令菜单,即可进入学生成绩统计界面,如图 4-11 所示。

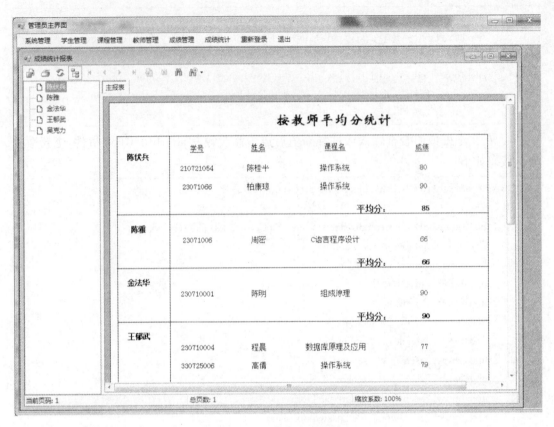

图 4-11 成绩统计界面

该窗体中主要用了一个 crystalReportView 控件,设置该控件的 ReportSource 属性为已经预先准备好的 crystal 报表——gradetj.rpt。

4.2.6.1 生成并设置 gradetj.rpt

依次点击"项目"/"添加组件",在左侧的类别中选择"Reportiong",在右侧已安装模板中选择"Crystal 报表",在名称中输入 gradetj.rpt,点击添加就生成了一个水晶报表。如下图 4-12 所示。

点击添加后,进入水晶报表设计界面如图 4-13 所示。

在字段资源管理器中,右击数据库字段,点击数据库专家,选择创建新连接下的 OLE DB 可以选择数据库类型,连接数据库,操作步骤如图 4-14、4-15 所示。

第4章 学生成绩管理系统

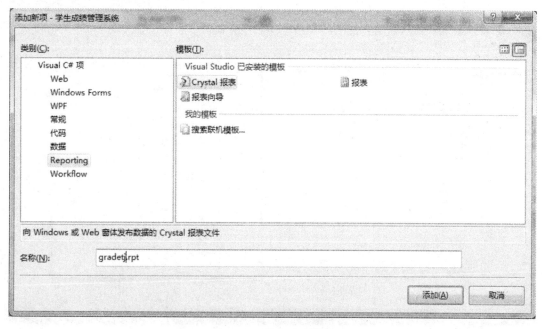

图 4-12 添加水晶报表

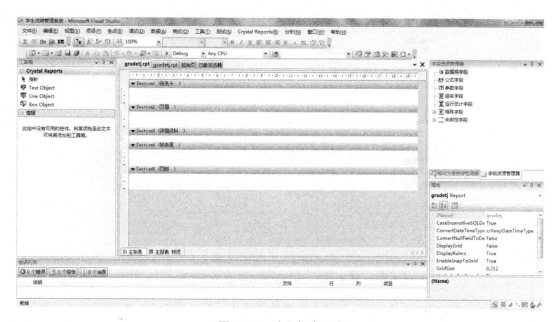

图 4-13 水晶报表设计

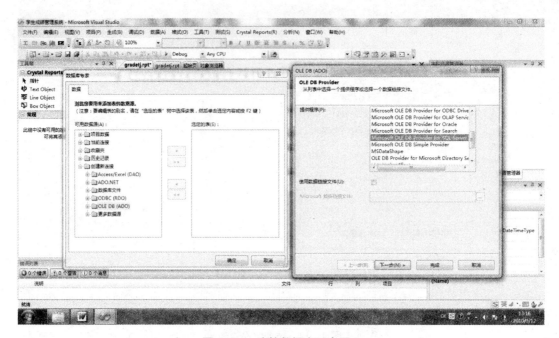

图 4-14 连接数据库示意图 1

图 4-15 连接数据库示意图 2

点击完成后如图 4-16 所示,可以选择水晶报表中需要的数据库中的数据表。

该步骤完成后,回到字段资源管理器中,就出现了数据表中各个字段,如图 4-17 所示。把需要的字段拖入主报表的详细资料区即可。若需要分组汇总,则需要插入分组;美化报表

第 4 章　学生成绩管理系统

的过程中可以给报表插入线、框等等,这些步骤请读者自行完成。至此,水晶报表设置完成。

图 4-16　选择数据表

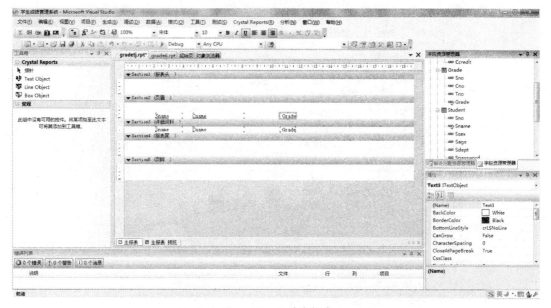

图 4-17　设计主报表

4.2.7 教师身份主界面设计

选择管理员身份登录后进入管理员主界面,如下图 4-18 所示。在该界面中设计了一个 menuStrip 控件,并且设置窗体的 IsMdiContainer 属性为 True。

该主界面的作用就是显示教师所拥有的所有的功能菜单项,当用户单击相应的菜单项时,打开对应的模块窗口。

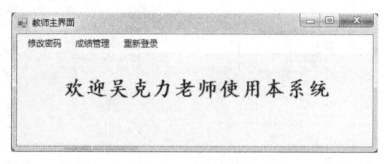

图 4-18 教师主界面

修改密码、重新登录实现方法同管理员主界面,这里不再赘述,请读者自行参考上文的代码。

在教师主界面中选择"成绩管理"命令菜单,即可进入教师成绩查询录入修改界面,如图 4-19 所示。

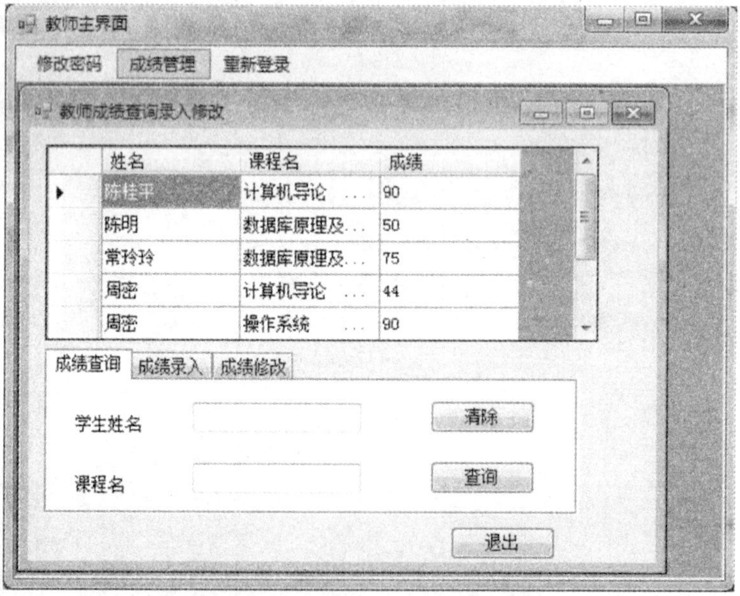

图 4-19 教师成绩查询录界面入修改

该窗体中主要了设计了 4 个 TextBox 控件、4 个 ComboBox 控件、5 个 Button 控件、1 个 TabControl 控件和 1 个 DataGridView 控件。各个控件的名称、作用如表 4-13 所示。

表 4-13 成绩查询录入修改界面控件设计

控件类型	控件名称	作用
TextBox	selectsnamebox	查询时输入的学生姓名
	selectcnamebox	查询时输入的课程名称
	insertgradebox	插入时输入的成绩
	updategradebox	修改时输入的成绩
Button	setnull	清除文本框内容
	chaxun	查询确认按钮
	extbtn	退出按钮
	okbtn	插入确认按钮
	updatebtn	更新确认按钮
ComboBox	insertsnamebox	插入时选择学生姓名
	insertcnamebox	插入时选择课程名称
	updatesnamebox	修改时显示学生姓名（Enable 属性设置为 False）
	updatecnamebox	修改时显示课程名称（Enable 属性设置为 False）
DataGridView	dataGridView1	显示教师所教课程成绩信息
TabControl	tabControl1	容器（设置 TabPages 属性）

1. 窗体加载时触发 teacherselectgrade_Load 事件，dataGridView1 中显示该教师所教课程名称和对应学生的成绩，在 tabPage2 的两个 ComboBox 中分别绑定学生姓名和课程名，并能获取到学生姓名所对应的学号，以及课程名所对应的课程号。代码如下：

```
private void teacherselectgrade_Load(object sender, EventArgs e)
{
    cn.Open();
    string str = string.Format("select student.sname,course.cname,grade.grade from student,course,teacher,grade where student.sno=grade.sno and course.cno=grade.cno and teacher.tno=grade.tno and teacher.tno='{0}'", Login.userid);
    SqlCommand cmd = new SqlCommand(str, cn);
```

```
    myadapter = new SqlDataAdapter(cmd);
    ds.Clear();
    myadapter.Fill(ds,"grade");
    dataGridView1.DataSource = ds.Tables[0];
    string str2 = "select sno,sname from student ";
    SqlDataAdapter adp = new SqlDataAdapter(str2,cn);
    adp.Fill(ds,"student");
    insertsnamebox.DataSource = ds.Tables["student"].DefaultView;
    insertsnamebox.DisplayMember = "sname";
    insertsnamebox.ValueMember = "sno";
    cn.Close();
    string str3 = "select cno,cname from course ";
    SqlDataAdapter adpt = new SqlDataAdapter(str3,cn);
    adpt.Fill(ds,"course");
    insertcnamebox.DataSource = ds.Tables["course"].DefaultView;
    insertcnamebox.DisplayMember = "cname";
    insertcnamebox.ValueMember = "cno";
    cn.Close();
}
```

2. 单击 dataGridView1 中的单元格触发 dataGridView1_CellClick 事件,将当前选择的行信息在 tabPage3 中的对应的文本框或组合框显示。代码如下:

```
private void dataGridView1_CellClick(object sender, DataGridViewCellEventArgs e)
{
    n = e.RowIndex;
    updatesnamebox.Text=dataGridView1.Rows[e.RowIndex].Cells[0].Value.ToString();
    updatecnamebox.Text=dataGridView1.Rows[e.RowIndex].Cells[1].Value.ToString();
    updategradebox.Text = dataGridView1.Rows[e.RowIndex].Cells[2].Value.ToString();
    updatesnamebox.DataSource = ds.Tables["student"].DefaultView;
    updatesnamebox.DisplayMember = "sname";
    updatesnamebox.ValueMember = "sno";
    updatecnamebox.DataSource = ds.Tables["course"].DefaultView;
    updatecnamebox.DisplayMember = "cname";
    updatecnamebox.ValueMember = "cno";
}
```

3. 单击 tabPage1 中的查询按钮触发 chaxun_Click 事件,实现按姓名模糊查询、按课程名模糊查询,或同时进行模糊查询。代码如下:

```csharp
private void chaxun_Click(object sender, EventArgs e)
{
    DataSet ds2 = new DataSet();
    cn.Open();
    string str = string.Format("select student.sname,course.cname,grade.grade from student,course,teacher,grade where student.sno=grade.sno and course.cno=grade.cno and teacher.tno=grade.tno and teacher.tno='{0}'", Login.userid);
    if (selectsnamebox.Text != null)
        str += string.Format(" and student.sname like '%{0}%'", selectsnamebox.Text);
    if (selectcnamebox.Text != null)
        str += string.Format(" and course.cname like '%{0}%'", selectcnamebox.Text);
    SqlCommand cmd = new SqlCommand(str, cn);
    myadapter = new SqlDataAdapter(cmd);
    ds2.Clear();
    myadapter.Fill(ds2, "grade");
    dataGridView1.DataSource = ds2.Tables[0];
    cn.Close();
}
```

4. 单击 tabPage2 中的确定按钮触发 okbtn_Click 事件,可以实现向 grade 表中添加记录。代码如下:

```csharp
private void okbtn_Click(object sender, EventArgs e)
{
    if (insertsnamebox.Text.Trim() == "" || insertcnamebox.Text.Trim() == "")
        MessageBox.Show("请输入完整信息");
    else
    {
        cn.Open();
        string sql = string.Format("select * from grade where sno='{0}' and cno='{1}' and tno='{2}'", insertsnamebox.SelectedValue.ToString(), insertcnamebox.SelectedValue.ToString(), Login.userid);
        SqlCommand cmd=new SqlCommand(sql,cn);
        if(null==cmd.ExecuteScalar())
        {
```

```
            string str = string.Format("insert into grade(sno,cno,tno,grade) values('{0}','{1}','{2}',{3})",
insertsnamebox.SelectedValue.ToString(),insertcnamebox.SelectedValue.ToString(),Login.userid,
insertgradebox.Text.Trim());
            cmd.CommandText = str;
            cmd.ExecuteNonQuery();
            MessageBox.Show("录入成功");
             string str2 = string.Format("select student.sname,course.cname,grade.grade from
student,course,teacher,grade where student.sno=grade.sno and course.cno=grade.cno and teacher.
tno=grade.tno and teacher.tno='{0}'",Login.userid);
            cmd.CommandText = str2;
            SqlDataAdapter myadapter = new SqlDataAdapter(cmd);
            ds.Clear();
            myadapter.Fill(ds,"grade");
            dataGridView1.DataSource = ds.Tables[0];
           }
       }
  }
```

5. 单击 tabPage3 中的确定按钮触发 updatebtn_Click 事件,实现对成绩的修改。代码如下:

```
       private void updatebtn_Click(object sender, EventArgs e)
       {
           string updatestring = string.Format("update grade set grade={0} where sno='{1}' and cno=
'{2}' and tno='{3}'",updategradebox.Text.Trim(),updatesnamebox.SelectedValue.ToString(),
updatecnamebox.SelectedValue.ToString(),Login.userid);
           SqlCommand cmd = new SqlCommand(updatestring,cn);
           if(cn.State==ConnectionState.Closed) cn.Open();
           cmd.ExecuteNonQuery();
           MessageBox.Show("更新成功");
           cn.Close();
           cn.Open();
           string str = string.Format("select student.sname,course.cname,grade.grade from student,course,teacher,
grade where student.sno=grade.sno and course.cno=grade.cno and teacher.tno=grade.tno and teacher.tno='{0}'",
Login.userid);
           cmd.CommandText = str;
           myadapter = new SqlDataAdapter(cmd);
```

```
        ds.Clear();
        myadapter.Fill(ds,"grade");
        dataGridView1.DataSource = ds.Tables[0];
        string str2 = "select sno,sname from student ";
        SqlDataAdapter adp = new SqlDataAdapter(str2,cn);
        adp.Fill(ds,"student");
        insertsnamebox.DataSource = ds.Tables["student"].DefaultView;
        insertsnamebox.DisplayMember = "sname";
        insertsnamebox.ValueMember = "sno";
        cn.Close();
        string str3 = "select cno,cname from course ";
        SqlDataAdapter adpt = new SqlDataAdapter(str3,cn);
        adpt.Fill(ds,"course");
        insertcnamebox.DataSource = ds.Tables["course"].DefaultView;
        insertcnamebox.DisplayMember = "cname";
        insertcnamebox.ValueMember = "cno";
        cn.Close();
    }
```

第 5 章 课程设计案例

数据库课程设计是培养学生应用所学基础理论和软件技术解决实际问题能力的重要教学环节。数据库应用系统的设计通常经历需求分析、系统设计(库表、功能模块)、系统实现(编码、调试)、测试几个阶段。本章通过对学生宿舍管理系统课程设计过程的剖析，较详细地介绍了数据库课程设计的主要内容、过程和实现方法。

5.1 数据库应用软件设计步骤

数据库及其应用系统开发全过程分为如下 6 个阶段：

1. 需求分析阶段

了解与分析用户需求(包括数据与处理)，是整个设计过程的基础，也是最困难、最耗费时间的一步。

2. 概念结构设计阶段

是整个数据库设计的关键，通过对用户需求进行综合、归纳与抽象，形成一个独立于具体 DBMS 的概念模型。通常用 E-R 图与用户进行交互、讨论。

3. 逻辑结构设计阶段

将概念结构转换为某个 DBMS 所支持的数据模型，对其进行优化。

4. 物理结构设计阶段

为逻辑数据模型选取一个最适合应用环境的物理结构(包括存储结构和存取方法)。

5. 数据库实施阶段

运用 DBMS 提供的数据语言、工具及宿主语言，根据逻辑设计和物理设计的结果建立数据库，编制与调试应用程序，组织数据入库，并进行试运行。

6. 数据库运行和维护阶段

数据库应用系统经过测试运行后可投入正式运行。

5.2 需求分析

需求分析的任务是通过详细调查现实世界要处理的对象，充分了解原系统(手工系统)工作概况，明确用户的各种需求，然后在此基础上确定新系统的功能。需求分析的重点是获得用户对数据库的信息要求、处理要求、安全性和完整性要求。

5.2.1 项目需求

本系统具有以下特点：

1. 实现宿舍信息、学生入住、宿舍报修、水电费缴存、外来人员登记的管理；
2. 满足宿舍管理员对宿舍管理的不同操作要求；
3. 界面设计简单、操作方便。

本系统后台数据库采用 SQL Server 2008，前台采用 Visual Studio C♯2010 作为主要开发工具。

管理员能查询宿舍楼的所有相关信息，包括：学生在宿舍楼住宿的详细信息、各宿舍卫生检查成绩信息、各宿舍水电缴费情况信息、各宿舍房屋报修信息、外来人员详细信息等。学生能查询本宿舍的所有相关信息，包括水电缴费情况、卫生检查成绩、宿舍报修记录等。

5.2.2 系统功能需求

对于该系统的管理员，需要具有如下信息处理功能：

1. 宿舍管理：查询宿舍详细信息、添加宿舍、修改宿舍信息（如电话号码变更等）；
2. 入住管理：学生宿舍的安排、宿舍的调整；
3. 卫生检查：卫生成绩查询、添加；
4. 水电收费：水电抄表、水电收缴；
5. 宿舍报修：报修登记、报修查询；
6. 外来人员登记：外来人员登记、查询。

依据信息及处理需求，可得到如下图 5-1 所示功能结构图。

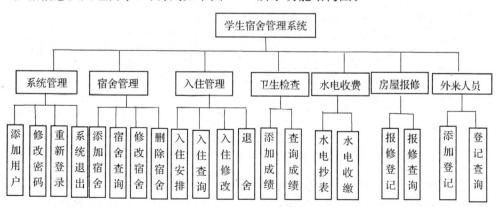

图 5-1 系统功能结构图

5.2.3 安全性、完整性要求

数据库的安全性是指保护数据库以防止不合法的使用所造成的数据泄露、更改或破坏。本系统的安全性要求如下：

1. 系统应设置用户的标识以鉴定是否为合法用户，并要求合法用户设置其密码，以防止身份被盗用；

2. 本系统主要为宿舍管理员使用，故只有一种权限。

数据库的完整性是指数据的正确性、相容性。本系统为了防止数据库中存在不符合语义的数据，做如下完整性要求：

1. 各种信息记录的完整性，信息记录内容不能为空；

2. 相同数据在不同记录中的一致性。

5.3 概念结构设计

概念结构设计就是将需求分析得到的用户需求抽象为信息结构即概念模型的过程。它是整个数据库设计的关键。

依据5.2的需求分析，抽象得到系统的概念模型如图5-2所示，每个实体及属性如下。

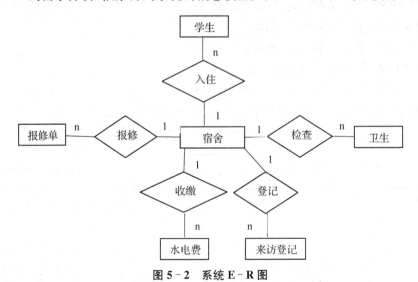

图 5-2 系统 E-R 图

- 学生：学号、姓名、性别、学院、班级
- 宿舍：宿舍号、宿舍床位数、宿舍桌子数、椅子数、宿舍朝向、住宿费、备注
- 来访登记：访客姓名、被访学生姓名、到访时间、离开时间

- 卫生检查:检查时间、检查成绩、备注
- 报修单:报修人、报修时间、报修备注
- 水电费:缴费类型、缴费月份、缴费金额、缴费人、缴费日期、缴费标记

5.4 数据库逻辑结构设计

逻辑结构设计的任务就是把概念结构阶段设计好的基本 E-R 图转换为与选用 DBMS 产品所支持的数据模型相符合的逻辑结构。本系统采用 SQL Server 数据库管理系统,即关系数据库管理系统,该阶段的任务是将 E-R 图转换为关系模型,并设计子模式的结构。

E-R 图转换为关系模式遵循如下原则:

1. 一个实体型转换为一个关系模式;
2. 一个 1∶1 联系可以转换为一个独立的关系模式,也可以与任意一端对应的关系模式合并;
3. 一个 1∶n 联系可以转换为一个独立的关系模式,也可以与 n 端对应的关系模式合并;
4. 一个 m∶n 联系转换为一个独立的关系模式。

5.4.1 关系模式设计

根据以上转换原则,E-R 图中的每个实体均转换成关系模式,1∶n 联系均与 n 端合并,并且在 n 端关系模式中加上 1 端的候选码,据此得到如下关系模式:

- 宿舍表 dorm(宿舍号、宿舍床位数、宿舍椅子数、宿舍朝向、住宿费、备注);
- 学生表 student(学号、姓名、性别、学院、班级,宿舍号);
- 报修表 repair(报修人、报修时间、报修备注,宿舍号);
- 水电费表 charge(缴费类型、缴费月份、缴费金额、缴费人、缴费日期、缴费标记,宿舍号);
- 卫生检查表 checkinformation(检查时间、检查成绩、备注,宿舍号);
- 外来人员登记表 register(检查时间、检查成绩、备注,宿舍号);
- 为保证系统使用过程中的安全性,登录系统需身份验证,故设计用户信息表 UserInformation(用户名,密码);

各个表的数据结构如下表 5-1~5-7 所示。

表 5-1 用户信息表(UserInformation)结构

列名	数据类型	允许 Null 值
UserName	varchar(20)	
UserPassword	varchar(20)	✓

表 5-2 学生表(Student)结构

列名	数据类型	允许 Null 值
🔑 sno	varchar(10)	☐
sname	varchar(20)	☑
ssex	varchar(2)	☑
sdept	varchar(10)	☑
sclass	varchar(10)	☑
dormID	varchar(8)	☑
		☐

表 5-3 宿舍表(dorm)结构

列名	数据类型	允许 Null 值
🔑 dormID	varchar(8)	☐
dormMoney	decimal(8, 2)	☑
dormbed	int	☑
dormchair	int	☑
dormdesk	int	☑
dormdirection	char(2)	☑
dormremark	text	☑
		☐

表 5-4 报修表(repair)结构

列名	数据类型	允许 Null 值
▶ dormID	varchar(8)	☑
repairperson	varchar(8)	☑
addrepairdate	datetime	☑
repairremark	text	☑
		☐

表 5-5 水电费表(charge)结构

列名	数据类型	允许 Null 值
🔑 dormID	varchar(8)	☐
🔑 chargetype	varchar(4)	☐
🔑 chargemonth	varchar(4)	☐
chargemoney	decimal(8, 2)	☑
chargeperson	varchar(8)	☑
chargedate	datetime	☑
flag	bit	☑
		☐

表 5-6　卫生检查表(checkinformation)结构

列名	数据类型	允许 Null 值
dormID	varchar(8)	
CheckDate	datetime	
CheckState	varchar(5)	✓
CheckRemark	text	✓

表 5-7　外来人员登记表(register)结构

列名	数据类型	允许 Null 值
dormID	varchar(8)	
vistor	varchar(8)	
bevisted	varchar(8)	✓
vistdatetime	datetime	
leaverdatetime	datetime	✓
registerremark	text	✓

定义库、表结构的 SQL 语句如下：

```
create database DormMangement
use DormManagement
create table userinformation
(
    username varchar(18) primary key,
    userpassword varchar(10)
)

create table dorm
(
    dormID varchar(8) primary key,
    dormMoney decimal(8,2),
    dormbed int,
    dormchair int,
    dormdesk int,
    dormdirection char(2),
    dormremark text
)
```

```sql
create table student
(
    sno varchar(10) primary key,
    sname varchar(20),
    ssex varchar(2),
    sdept varchar(10),
    sclass varchar(10),
    dormID varchar(8),
    foreign key(dormID) references dorm(dormID)
)

create table checkinformation
(
    dormID varchar(8),
    CheckDate datetime,
    CheckState varchar(5),
    CheckRemark text,
    primary key(dormID,CheckDate),
    foreign key(dormID) references dorm(dormID)
)

create table charge
(
    dormID varchar(8),
    chargetype varchar(4),
    chargemonth varchar(4),
    chargemoney decimal(8,2),
    chargeparson varchar(8),
    chargedate datetime,
    primary key(dormID,chargetype,chargemonth),
    foreign key(dormID) references dorm(dormID)
)

create table repair
(
    dormID varchar(8),
    repairperson varchar(8),
```

```
    addrepairdate datetime,
    repairremark text,
    primary key(dormID,repairperson,addrepairdate),
    foreign key(dormID) references dorm(dormID)
)

create table register
(
    dormID varchar(8),
    vistor varchar(8),
    bevisted varchar(8),
    vistdatetime datetime,
    leaverdatetime datetime,
    registerremark text,
    primary key(dormID,vistor,vistdatetime),
    foreign key(dormID) references dorm(dormID)
)
```

表与表之间的关系如图 5-3 所示。

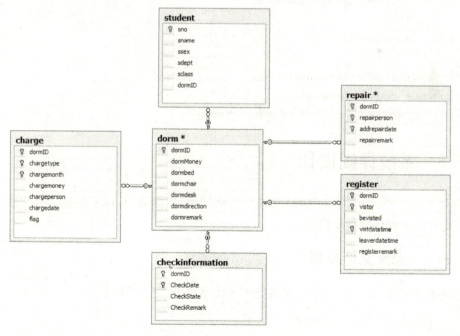

图 5-3　Dorm Management 关系图

5.4.2 子模式设计

将概念模型转换为全局逻辑模型后,还应该根据局部应用需求,结合具体的 DBMS 特点,设计用户的外模式。SQL Server 数据库管理系统提供了视图概念,可以利用这一功能设计更符合局部用户需求的用户外模式。本系统为方便统计各宿舍已入住人数及剩余床位数,设计如下视图:

```
create view numberofstudent(dormID,numberofstudent)
as
select dorm.dormID,COUNT(sno)
from dorm
left outer join student on dorm.dormid=student.dormid
group by dorm.dormID

create view numberofbed(dormID,numberofbed)
as
select dorm.dormID,dormbed-numberofstudent
from dorm,numberofstudent
where dorm.dormID=numberofstudent.dormID

Create View dormcount
as
select dormID,numberofdorm,numberofbed
from numberofbed,numberofdorm
where numberofdorm.dormID=numberofbed.dormID
```

5.5 系统详细设计

启动 Visual Studio 2010,选择"文件""新建""项目"即命令,建立学生宿舍管理应用程序。下面介绍各窗体的程序设计。

5.5.1 配置文件设置

新建配置文件 App.config,定义数据库连接字符串,定义方法如下:

```
<?xml version="1.0" encoding="utf-8"?>
<configuration>
```

```
        <configSections>
        </configSections>
        <connectionStrings>
            <add name="mycon" connectionString="data source=202.195.126.161;initial catalog=DormManagement;User ID=sa;password=123456" />
            <add name="DormManagement.Properties.Settings.DormManagementConnectionString"
                connectionString="Data Source=202.195.126.161;Initial Catalog=DormManagement;User ID=sa;password=123456"
                providerName="System.Data.SqlClient" />
        </connectionStrings>
    </configuration>
```

5.5.2 登录界面设计

登录界面的作用就是判断用户身份的合法性。输入用户名和密码,选择用户身份,若用户名、密码、验证码均正确,单击[确定]按钮,即可进入主界面,否则不能进入。登录界面效果,如图5-4所示。

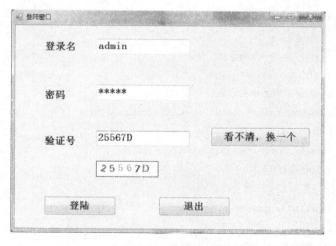

图 5-4 登录界面

该窗体中各个控件的名称、属性、事件如表 5-7 所示。

表 5-8 登录界面主要控件设计

控件类型	控件名称	属性值	事件	备注
TextBox	txtusername	Text="admin"		
	txtpassword	Text="admin" passwordchar="*"		
	txtcheckid	Text=""		
PictureBox	pictureBox1			
Button	btnlogin	Text=确定	Click	
	btnexit	Text=退出	Click	
	btnnext	Text=看不清,换一个	Click	

登录界面部分代码:

```
using System.Configuration;
using System.Data.SqlClient;

namespace DormManagement
{
    public partial class LoginFrm : Form
    {
        private const int codelength = 6;
        private string randomcode = "";
        public static string username = "";
        public static string constr = ConfigurationManager.ConnectionStrings["mycon"].ConnectionString.ToString();
        SqlConnection cn = new SqlConnection(constr);
        public LoginFrm()
        {
            InitializeComponent();
        }
        private string CreateRandomeCode(int length)
        {
            //生成长度为 length 的随机字符串
            int rand;
            char code;
            string randomcode = string.Empty;
```

```csharp
            System.Random random = new Random();
            for (int i = 0; i < length; i++)
            {
                rand = random.Next();
                if (rand % 3 == 0)
                    code = (char)('A' + (char)(rand % 26));
                else
                {
                    code = (char)('0' + (char)(rand % 10));
                }
                randomcode += code.ToString();
            }
            return randomcode;
        }

        private void createimage(string randomcode)
        {
            if (randomcode == null || randomcode.Trim() == String.Empty)
            {
                return;
            }
            System.Drawing.Bitmap image = new Bitmap((int)Math.Ceiling(randomcode.Length * 20.0), 30);
            Graphics g = Graphics.FromImage(image);
            try
            {
                //绘制边框
                int randAngle = 30;
                g.Clear(Color.White);
                g.DrawRectangle(new Pen(Color.Black, 0), 0, 0, image.Width - 1, image.Height - 1);
                g.SmoothingMode = System.Drawing.Drawing2D.SmoothingMode.AntiAlias;
                Random rand = new Random();

                //背景噪点生成
```

```csharp
Pen blackPen = new Pen(Color.LightBlue, 0);
for (int i = 0; i < 50; i++)
{
    int x = rand.Next(0, pictureBox1.Width);
    int y = rand.Next(0, pictureBox1.Height);
    g.DrawRectangle(blackPen, x, y, 1, 1);
}
char[] chars = randomcode.ToCharArray();
//拆散字符串成单个字符数组

//定义文字居中
StringFormat format = new StringFormat(StringFormatFlags.NoClip);
format.Alignment = StringAlignment.Center;
format.LineAlignment = StringAlignment.Center;

//定义颜色
Color[] c = { Color.Black, Color.Red, Color.LimeGreen, Color.MidnightBlue, Color.Green, Color.Blue };
//定义字体
string[] font = { "Microsoft Sans Serif", "Arial", "宋体" };

for (int i = 0; i < chars.Length; i++)
{
    int cindex = rand.Next(6);
    int findex = rand.Next(3);
    Font f = new Font(font[findex], 14, System.Drawing.FontStyle.Bold);
    Brush b = new System.Drawing.SolidBrush(c[cindex]);
    Point dot = new Point(14, 14);
    float angle = rand.Next(-randAngle, randAngle);
    g.TranslateTransform(dot.X, dot.Y);
    g.RotateTransform(angle);
    g.DrawString(chars[i].ToString(), f, b, 1, 1, format);
    g.RotateTransform(-angle);
    g.TranslateTransform(2, -dot.Y);
}
this.pictureBox1.Width = image.Width;
```

```csharp
            this.pictureBox1.Height = image.Height;
            pictureBox1.BackgroundImage = image;
        }
        catch (ArgumentException)
        {
            MessageBox.Show("创建图片错误");
        }
    }

    private void btnexit_Click(object sender, EventArgs e)
    {
        Application.Exit();
    }

    private void btnlogin_Click(object sender, EventArgs e)
    {
        //登录按钮
        string str = string.Format("select * from userinformation where username='{0}' and userpassword='{1}'", txtusername.Text, txtpassword.Text);
        SqlCommand cmd = new SqlCommand(str, cn);
        if (txtusername.Text == "")
        {
            MessageBox.Show("用户名不能为空");
        }
        else if (txtpassword.Text == "")
        {
            MessageBox.Show("密码不能为空");
        }
        else if (txtcheckid.Text == "")
        {
            MessageBox.Show("请输入验证码");
        }
        else
        {
            try
```

```
            {
                cn.Open();
                SqlDataReader dr = cmd.ExecuteReader();
                if (dr.Read())
                {
                    if (txtcheckid.Text == randomcode)
                    {
                        MainFrm frm = new MainFrm();
                        username = txtusername.Text;
                        frm.Show();
                        this.Hide();
                    }
                    else
                    {
                        MessageBox.Show("验证码错误,请重新输入");
                        txtcheckid.Focus();
                        txtcheckid.SelectAll();
                    }
                }
                else
                {
                    MessageBox.Show("用户名或密码错误,请重新输入");
                    txtusername.Focus();
                    txtusername.SelectAll();
                }
            }
            catch (SqlException ex)
            {
                MessageBox.Show(ex.Message.ToString());
            }
            finally
            {
                cn.Close();
            }
        }
    }
```

```
            private void btnnext_Click(object sender, EventArgs e)
            {
                randomcode = CreateRandomeCode(codelength);
                createimage(randomcode);
            }

            private void Form1_Load(object sender, EventArgs e)
            {
                randomcode = CreateRandomeCode(codelength);
                createimage(randomcode);
            }
        }
```

5.5.3 主界面设计

输入正确的用户名、密码及验证码之后,进入系统主界面,主界面及各菜单下拉菜单如下系列图 5-5 所示。

图 5-5 主窗体界面

该窗体中各个控件的名称、属性、事件如表 5-9 所示，各菜单、工具栏均触发 Click 事件。

表 5-9　主窗体主要控件设计

控件类型	控件名称	属性值	事件	备注
Form	MainForm	Text="主窗体"		
TextBox	txtusername	Text="admin"		
	txtpassword	Text="admin" passwordchar="*"		
	txtcheckid	Text=""		
MenuStrip	menuStrip1		Click	具体菜单内容如上图所示
ToolStrip	toolStrip1		Click	工具栏内容设置如上图所示
Button	btnlogin	Text=确定	Click	
	btnexit	Text=退出	Click	
	btnnext	Text=看不清，换一个	Click	

主界面代码：

```
private void 添加用户 ToolStripMenuItem_Click(object sender, EventArgs e)
{
    AddUser addfrm = new AddUser();
    for (int x = 0; x < this.MdiChildren.Length; x++)
    {
        Form tempchildform = this.MdiChildren[x];
        tempchildform.Close();
    }
    addfrm.MdiParent = this;
    addfrm.WindowState = FormWindowState.Maximized;
    addfrm.Show();
}

private void 修改密码 ToolStripMenuItem_Click(object sender, EventArgs e)
{
    Modifypassword modifyFrm = new Modifypassword();
    for (int x = 0; x < this.MdiChildren.Length; x++)
    {
```

```
                Form tempchildform = this.MdiChildren[x];
                tempchildform.Close();
            }
            modifyFrm.MdiParent = this;
            modifyFrm.WindowState = FormWindowState.Maximized;
            modifyFrm.Show();
        }

        private void 重新登录ToolStripMenuItem_Click(object sender, EventArgs e)
        {
            LoginFrm f1 = new LoginFrm();
            f1.Show();
            this.Close();
        }

        private void 退出ToolStripMenuItem_Click(object sender, EventArgs e)
        {
            this.Close();
        }

        private void 添加宿舍ToolStripMenuItem_Click(object sender, EventArgs e)
        {
            AddDorm adddorm = new AddDorm();
            for (int i = 0; i < this.MdiChildren.Length; i++)
            {
                Form tempchild = (Form)this.MdiChildren[i];
                tempchild.Close();
            }
            adddorm.MdiParent = this;
            adddorm.WindowState = FormWindowState.Maximized;
            adddorm.Show();
        }

        private void 查询宿舍ToolStripMenuItem_Click(object sender, EventArgs e)
        {
            SearchDorm searchdorm = new SearchDorm();
            for (int i = 0; i < this.MdiChildren.Length; i++)
```

```csharp
        {
            Form tempchild = (Form)this.MdiChildren[i];
            tempchild.Close();
        }
        searchdorm.MdiParent = this;
        searchdorm.WindowState = FormWindowState.Maximized;
        searchdorm.Show();
    }

    private void 修改宿舍信息ToolStripMenuItem_Click(object sender, EventArgs e)
    {
        SearchDorm searchdorm = new SearchDorm();
        for (int i = 0; i < this.MdiChildren.Length; i++)
        {
            Form tempchild = (Form)this.MdiChildren[i];
            tempchild.Close();
        }
        searchdorm.MdiParent = this;
        searchdorm.WindowState = FormWindowState.Maximized;
        searchdorm.Show();
    }

    private void 删除宿舍ToolStripMenuItem_Click(object sender, EventArgs e)
    {
        SearchDorm searchdorm = new SearchDorm();
        for (int i = 0; i < this.MdiChildren.Length; i++)
        {
            Form tempchild = (Form)this.MdiChildren[i];
            tempchild.Close();
        }
        searchdorm.MdiParent = this;
        searchdorm.WindowState = FormWindowState.Maximized;
        searchdorm.Show();
    }

    private void 学生入住安排ToolStripMenuItem_Click(object sender, EventArgs e)
    {
```

```csharp
        AddStudent addstudent = new AddStudent();
        for (int i = 0; i < this.MdiChildren.Length; i++)
        {
            Form tempchild = (Form)this.MdiChildren[i];
            tempchild.Close();
        }
        addstudent.MdiParent = this;
        addstudent.WindowState = FormWindowState.Maximized;
        addstudent.Show();
}

private void 学生查询ToolStripMenuItem_Click(object sender, EventArgs e)
{
        SearchStudent searchstudent = new SearchStudent();
        for (int i = 0; i < this.MdiChildren.Length; i++)
        {
            Form tempchild = (Form)this.MdiChildren[i];
            tempchild.Close();
        }
        searchstudent.MdiParent = this;
        searchstudent.WindowState = FormWindowState.Maximized;
        searchstudent.Show();
}

private void 添加成绩ToolStripMenuItem_Click(object sender, EventArgs e)
{
        AddCheck addcheck = new AddCheck();
        for (int i = 0; i < this.MdiChildren.Length; i++)
        {
            Form tempchild = (Form)this.MdiChildren[i];
            tempchild.Close();
        }
        addcheck.MdiParent = this;
        addcheck.WindowState = FormWindowState.Maximized;
        addcheck.Show();
}
```

```csharp
private void 查询成绩ToolStripMenuItem_Click(object sender, EventArgs e)
{
    SearchCheck searchcheck = new SearchCheck();
    for (int i = 0; i < this.MdiChildren.Length; i++)
    {
        Form tempchild = (Form)this.MdiChildren[i];
        tempchild.Close();
    }
    searchcheck.MdiParent = this;
    searchcheck.WindowState = FormWindowState.Maximized;
    searchcheck.Show();
}

private void 水电抄表ToolStripMenuItem_Click(object sender, EventArgs e)
{
    AddRecord addrecord = new AddRecord();
    for (int i = 0; i < this.MdiChildren.Length; i++)
    {
        Form tempchild = (Form)this.MdiChildren[i];
        tempchild.Close();
    }
    addrecord.MdiParent = this;
    addrecord.WindowState = FormWindowState.Maximized;
    addrecord.Show();
}

private void 水电收缴ToolStripMenuItem_Click(object sender, EventArgs e)
{
    AddCharge addcharge = new AddCharge();
    for (int i = 0; i < this.MdiChildren.Length; i++)
    {
        Form tempchild = (Form)this.MdiChildren[i];
        tempchild.Close();
    }
    addcharge.MdiParent = this;
    addcharge.WindowState = FormWindowState.Maximized;
    addcharge.Show();
```

}

private void 报修登记ToolStripMenuItem_Click(object sender, EventArgs e)
{
    AddRepair addrepair = new AddRepair();
    for (int i = 0; i < this.MdiChildren.Length; i++)
    {
        Form tempchild = (Form)this.MdiChildren[i];
        tempchild.Close();
    }
    addrepair.MdiParent = this;
    addrepair.WindowState = FormWindowState.Maximized;
    addrepair.Show();
}

private void 报修查询ToolStripMenuItem_Click(object sender, EventArgs e)
{
    SearchRepair searchrepair = new SearchRepair();
    for (int i = 0; i < this.MdiChildren.Length; i++)
    {
        Form tempchild = (Form)this.MdiChildren[i];
        tempchild.Close();
    }
    searchrepair.MdiParent = this;
    searchrepair.WindowState = FormWindowState.Maximized;
    searchrepair.Show();
}

private void 添加登记ToolStripMenuItem_Click(object sender, EventArgs e)
{
    AddRegister addRegister = new AddRegister();
    for (int i = 0; i < this.MdiChildren.Length; i++)
    {
        Form tempchild = (Form)this.MdiChildren[i];
        tempchild.Close();
    }
    addRegister.MdiParent = this;

```
        addRegister.WindowState = FormWindowState.Maximized;
        addRegister.Show();
    }

    private void 登记查询ToolStripMenuItem_Click(object sender, EventArgs e)
    {
        SearchRegister searchRegister = new SearchRegister();
        for (int i = 0; i < this.MdiChildren.Length; i++)
        {
            Form tempchild = (Form)this.MdiChildren[i];
            tempchild.Close();
        }
        searchRegister.MdiParent = this;
        searchRegister.WindowState = FormWindowState.Maximized;
        searchRegister.Show();
    }

    private void 宿舍入住统计报表ToolStripMenuItem_Click(object sender, EventArgs e)
    {
        DormCount dormcount = new DormCount();
        for (int i = 0; i < this.MdiChildren.Length; i++)
        {
            Form tempchild = (Form)this.MdiChildren[i];
            tempchild.Close();
        }
        dormcount.MdiParent = this;
        dormcount.WindowState = FormWindowState.Maximized;
        dormcount.Show();
    }
```

5.5.4 系统管理

5.5.4.1 添加用户

在主界面中选择"系统管理"/"添加用户"命令菜单,即可进入添加用户界面,如图 5-6 所示。该窗体中需要提供的信息包括:管理员用户名、密码、确认密码。信息录入后单击"确定"按钮,如果新用户名不存在并且密码和确认密码相同则向数据库中插入一条信息。

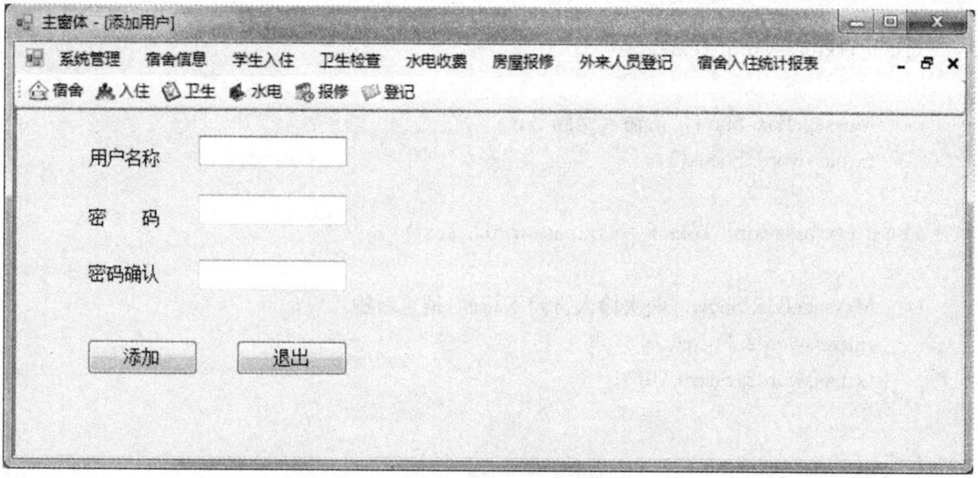

图 5-6 添加用户界面

在该窗体中各个控件的名称、属性、事件如表 5-10 所示。

表 5-10 添加管理员界面控件设计

控件类型	控件名称	属性值	事件	备注
Form	addUser	Text="添加用户"		
TextBox	txtusername			用户名
	txtpassword			用户密码
	txtpassword2			确认密码
Button	btnadd	Text=确定	Click	
	btnexit	Text=退出	Click	

用户界面代码：

```
SqlConnection cn = new SqlConnection(LoginFrm.constr);
private void btnadd_Click(object sender, EventArgs e)
{
    string insertstring = string.Format("insert into userinformation values('{0}','{1}')", txtusername.Text, txtpassword.Text);
    if (txtusername.Text == "")
    {
        MessageBox.Show("用户名不能为空,请输入用户名");
        txtusername.Focus();
```

```csharp
        }
        else if (txtpassword.Text == "")
        {
            MessageBox.Show("请输入密码");
            txtpassword.Focus();
        }
        else if (txtpassword.Text != txtpassword2.Text)
        {
            MessageBox.Show("两次输入密码不相同,请重新输入");
            txtpassword2.Focus();
            txtpassword2.SelectAll();
        }
        else
        {
            SqlCommand cmd = new SqlCommand(insertstring, cn);
            try
            {
                cn.Open();
                cmd.ExecuteNonQuery();
                MessageBox.Show("插入成功");
                txtpassword.Text = "";
                txtpassword2.Text = "";
                txtusername.Text = "";
            }
            catch (SqlException ex)
            {
                MessageBox.Show(ex.Message.ToString());
            }
            finally
            {
                cn.Close();
            }
        }
    }
    private void btnexit_Click(object sender, EventArgs e)
    {
        Application.Exit();
    }
```

5.5.4.2 修改密码

在主界面中选择"系统管理"/"修改密码"命令菜单,即可进入修改密码界面,如图 5-7 所示。每个管理员只能修改自己的密码,所以需要利用 Login 窗体中传递的公共变量 username,在 Load 事件中将该值传递给 txtusername 的 Text 属性,并且设置 txtusername 的 ReadOnly 属性为 True。

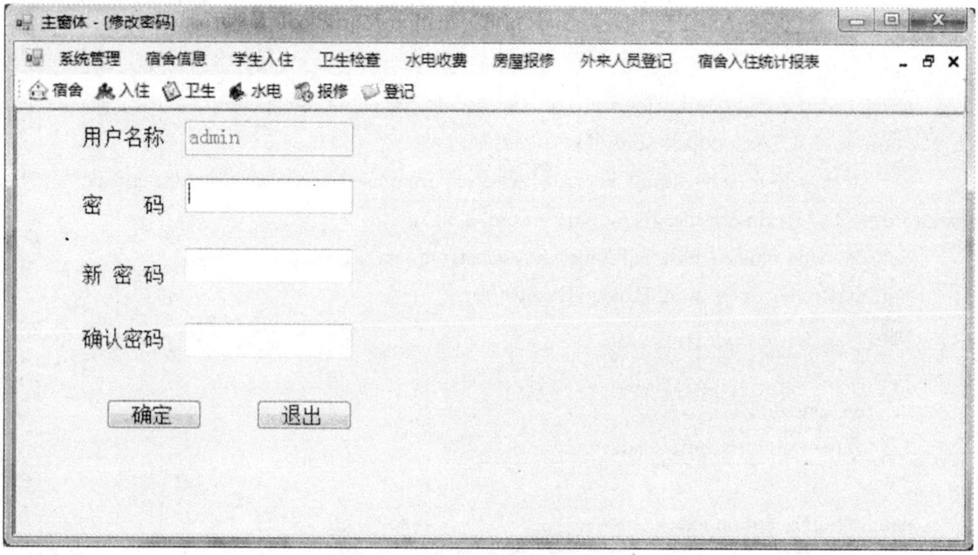

图 5-7 修改用户密码界面

在该窗体中各个控件的名称、属性、事件、备注如表 5-11 所示。

表 5-11 修改密码控件设计

控件类型	控件名称	属性值	事件	备注
Form	modifyUser	Text="修改密码"	Load	
TextBox	txtusername			用户名
	txtpassword			用户密码
	txtnewpassword			新密码
	txtnewpassword2			确认密码
Button	btnmodify	Text=确定	Click	
	btnexit	Text=退出	Click	

修改密码界面代码:

```csharp
SqlConnection cn = new SqlConnection(LoginFrm.constr);
public Modifypassword()
{
    InitializeComponent();
}

private void btnmodify_Click(object sender, EventArgs e)
{
    string updatestring = string.Format("update userinformation set userpassword='{0}' where username='{1}'", txtusername.Text, txtpassword.Text);
    string selestring = string.Format("select * from userinformation where username='{0}' and userpassword='{1}'", txtusername.Text, txtpassword.Text);
    SqlCommand cmd = new SqlCommand(selestring, cn);
    SqlDataReader dr = cmd.ExecuteReader();
    try
    {
        cn.Open();
        //dr=cmd.ExecuteReader();
    }
    catch (SqlException ex)
    {
        MessageBox.Show(ex.Message.ToString());
    }
    finally
    {
        cn.Close();
    }
    if (txtusername.Text == "")
    {
        MessageBox.Show("用户名不能为空,请输入用户名");
        txtusername.Focus();
    }
    else if (txtpassword.Text == "")
    {
        MessageBox.Show("请输入密码");
        txtpassword.Focus();
    }
```

```csharp
        else if (txtnewpassword.Text != txtnewpassword2.Text)
        {
            MessageBox.Show("两次输入密码不相同,请重新输入");
            txtnewpassword2.Focus();
            txtnewpassword2.SelectAll();
        }
        else if (dr.Read())
        {
            SqlCommand updatecmd = new SqlCommand(updatestring, cn);
            try
            {
                cn.Open();
                cmd.ExecuteNonQuery();
                MessageBox.Show("修改成功");
            }
            catch (SqlException ex)
            {
                MessageBox.Show(ex.Message.ToString());
            }
            finally
            {
                cn.Close();
            }
        }
    }

    private void Modifypassword_Load(object sender, EventArgs e)
    {
        txtusername.Text = LoginFrm.username;
        txtusername.Enabled = false;
    }

    private void btnexit_Click(object sender, EventArgs e)
    {
        this.Close();
    }
```

5.5.5 宿舍信息

5.5.5.1 添加宿舍

在主界面中选择"宿舍信息"/"添加宿舍"命令菜单,即可进入添加宿舍界面,如图 5-8 所示。

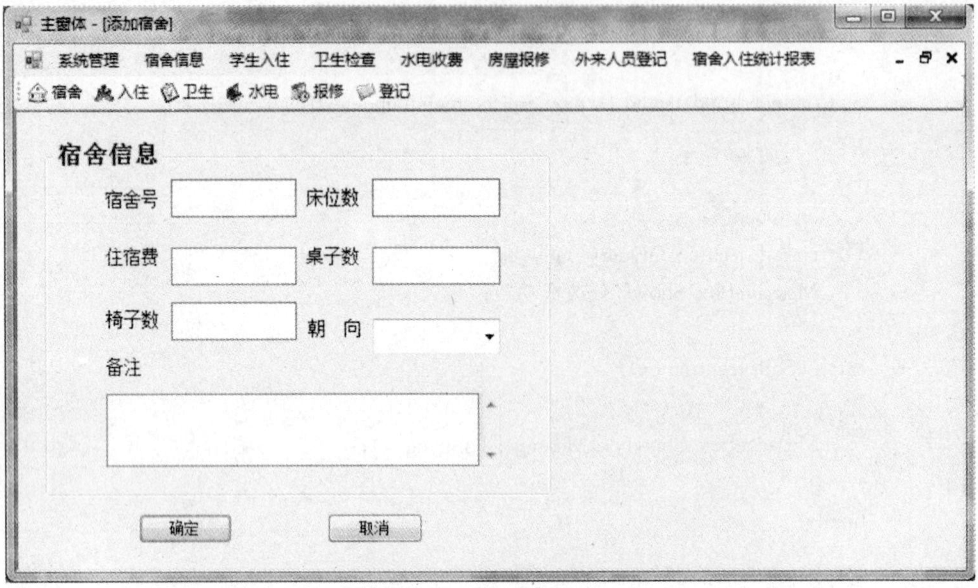

图 5-8 添加宿舍界面

该窗体中各个控件的名称、属性、事件、备注信息如表 5-12 所示。

表 5-12 添加宿舍界面控件设计

控件类型	控件名称	属性值	事件	备注
Form	AddDorm	Text="添加宿舍"		
TextBox	txtdormID			宿舍 ID
	txtdormbed			宿舍床位数
	txtdormMoney			住宿费
	txtdormdesk			宿舍桌子数
	txtdormdirection			宿舍朝向
	txtdormremark			

(续表)

控件类型	控件名称	属性值	事件	备注
Button	btnadddrom	Text=确定	Click	
	btnexit	Text=退出	Click	

添加宿舍界面主要代码：

```
private static string str = ConfigurationManager.ConnectionStrings["mycon"].ConnectionString.ToString();
SqlConnection cn=new SqlConnection(str);
private void btnadddorm_Click(object sender, EventArgs e)
{
    if(txtdormID.Text=="" || txtdormbed.Text=="")
        MessageBox.Show("请输入完整信息");
    else
    {
        try
        {
            if(cn.State==ConnectionState.Closed)
                cn.Open();
            string sqlstring = string.Format("select * from dorm where dormID='{0}'", txtdormID.Text);
            SqlCommand cmd=new SqlCommand(sqlstring,cn);
            if(cmd.ExecuteScalar()==null)
            {
                string insertsqlstring = string.Format("insert into dorm(dormid,dormMoney,dormbed,dormchair,dormdesk,dormdirection,dormremark) values ('{0}',{1},{2},{3},{4},'{5}','{6}')", txtdormID.Text, txtdormMoney.Text, txtdormbed.Text, txtdormchair.Text, txtdormdesk.Text, txtdormdirection.Text, txtdormremark.Text);
                cmd.CommandText=insertsqlstring;
                cmd.ExecuteNonQuery();
                MessageBox.Show("添加宿舍成功","提示信息");
                clear();
            }
            else
            {
                MessageBox.Show("宿舍号重复");
```

```
            }
        }
        catch(SqlException ex)
        {
            MessageBox.Show(ex.Message.ToString());
        }
        finally
        {
            if(cn.State==ConnectionState.Open)
                cn.Close();
        }
    }
}
private void clear()
{
    txtdormbed.Text="";
    txtdormchair.Text="";
    txtdormdesk.Text="";
    txtdormdirection.Text="";
    txtdormID.Text="";
    txtdormMoney.Text="";
    txtdormremark.Text="";
}
private void btncancel_Click(object sender, EventArgs e)
{
    this.Close();
}
```

5.5.5.2 宿舍查询

在主界面中选择"宿舍信息"/"查询宿舍"命令菜单,即可进入宿舍查询界面,如图5-9所示。

该窗体中各个控件的名称、属性、事件、备注信息如表5-13所示。注:该窗体调用了ModifyDorm窗体中的部分控件,所以,在ModifyDorm窗体中将这些控件的Modifiers属性设为Public。

第5章 课程设计案例

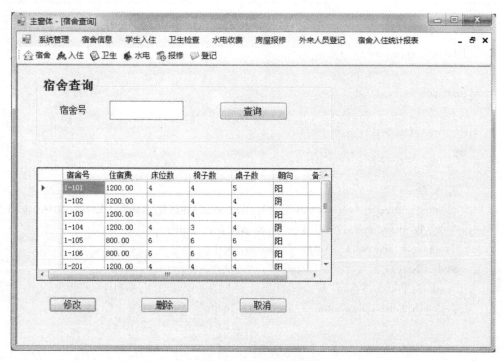

图5-9 宿舍查询界面

表5-13 添加宿舍界面控件设计

控件类型	控件名称	属性值	事件	备注
Form	searchDorm	Text="宿舍查询"		
TextBox	txtdormID			宿舍ID
DataGridView	dataGridview1			
Button	btnupdate	Text=修改	Click	
Button	btndelete	Text=删除	Click	
Button	btncancel	Text=取消	Click	

查询宿舍界面主要代码：

```
private DataSet ds;
private static string str = ConfigurationManager.ConnectionStrings["mycon"].ConnectionString.ToString();
SqlConnection cn=new SqlConnection(str);
SqlDataAdapter adpter;
```

```csharp
public SearchDorm()
{
    InitializeComponent();
}
void displaydb()
{
    ds = new DataSet();
    try
    {
        cn.Open();
        string sqlstring = "select dormID 宿舍号, dormMoney 住宿费, dormbed 床位数, dormchair 椅子数, dormdesk 桌子数, dormdirection 朝向, dormremark 备注 from dorm";
        adpter = new SqlDataAdapter(sqlstring, cn);
        ds.Clear();
        adpter.Fill(ds, "dorm");
        dataGridView1.DataSource = ds.Tables["dorm"];

    }
    catch (SqlException ex)
    {
        MessageBox.Show(ex.Message.ToString());
    }
    finally
    {
        cn.Close();
    }
}
private void btnDormSearch_Click(object sender, EventArgs e)
{
    ds = new DataSet();
    try
    {
        cn.Open();
        string sqlstring;
        if (txtdormID.Text == "")
```

```csharp
                sqlstring = "select dormID 宿舍号, dormMoney 住宿费, dormbed 床位数, dormchair 椅子数, dormdesk 桌子数, dormdirection 朝向, dormremark 备注 from dorm";
            else
                sqlstring = "select dormID 宿舍号, dormMoney 住宿费, dormbed 床位数, dormchair 椅子数, dormdesk 桌子数, dormdirection 朝向, dormremark 备注 from dorm where dormID='" + txtdormID.Text + "'";
            adpter = new SqlDataAdapter(sqlstring, cn);
            ds.Clear();
            adpter.Fill(ds, "dorm");
            dataGridView1.DataSource = ds.Tables["dorm"];

        }
        catch (SqlException ex)
        {
            MessageBox.Show(ex.Message.ToString());
        }
        finally
        {
            cn.Close();
            txtdormID.Text = "";
        }
    }
    ModifyDorm modifydorm;
    private void btnupdate_Click(object sender, EventArgs e)
    {
        if (dataGridView1.DataSource != null && dataGridView1.CurrentCell.RowIndex >= 0 && dataGridView1.CurrentCell != null)
        {
            modifydorm = new ModifyDorm();
            modifydorm.txtDormID.Text = ds.Tables["dorm"].Rows[dataGridView1.CurrentCell.RowIndex]["宿舍号"].ToString().Trim();
            modifydorm.txtDormBed.Text = ds.Tables["dorm"].Rows[dataGridView1.CurrentCell.RowIndex]["床位数"].ToString().Trim();
            modifydorm.txtDormChair.Text = ds.Tables["dorm"].Rows[dataGridView1.CurrentCell.RowIndex]["椅子数"].ToString().Trim();
```

```csharp
            modifydorm.txtDormDesk.Text = ds.Tables["dorm"].Rows[dataGridView1.CurrentCell.RowIndex]["桌子数"].ToString().Trim();
            modifydorm.txtDormDirection.Text = ds.Tables["dorm"].Rows[dataGridView1.CurrentCell.RowIndex]["朝向"].ToString().Trim();
            modifydorm.txtDormMoney.Text = ds.Tables["dorm"].Rows[dataGridView1.CurrentCell.RowIndex]["住宿费"].ToString().Trim();
            modifydorm.txtdormremark.Text = ds.Tables["dorm"].Rows[dataGridView1.CurrentCell.RowIndex]["备注"].ToString().Trim();
            modifydorm.ShowDialog();
        }
        displaydb();
    }

    private void btndelete_Click(object sender, EventArgs e)
    {
        SqlCommandBuilder cmdbuilder = new SqlCommandBuilder(adpter);
        if (MessageBox.Show("真的删除吗?", "提示", MessageBoxButtons.YesNoCancel, MessageBoxIcon.Warning) == DialogResult.Yes)
        {
            DataRow delrow = ds.Tables["dorm"].Rows[dataGridView1.CurrentCell.RowIndex];
            delrow.Delete();
            adpter.Update(ds, "dorm");
            MessageBox.Show("删除成功");
        }
        string sqlstring;
        if (txtdormID.Text == "")
            sqlstring = "select dormID 宿舍号, dormMoney 住宿费, dormbed 床位数, dormchair 椅子数, dormdesk 桌子数, dormdirection 朝向, dormremark 备注 from dorm";
        else
            sqlstring = "select dormID 宿舍号, dormMoney 住宿费, dormbed 床位数, dormchair 椅子数, dormdesk 桌子数, dormdirection 朝向, dormremark 备注 from dorm where dormID='" + txtdormID.Text + "'";
        adpter = new SqlDataAdapter(sqlstring, cn);
        ds.Clear();
        adpter.Fill(ds, "dorm");
        dataGridView1.DataSource = ds.Tables["dorm"];
```

```
    }
    private void btncancel_Click(object sender, EventArgs e)
    {
        this.Close();
    }
    private void SearchDorm_Load(object sender, EventArgs e)
    {
        displaydb();
    }
```

5.5.5.3 宿舍修改

在主界面中选择"宿舍信息"/"查询宿舍"命令菜单,即可进入宿舍查询界面,如图 5-9 所示,点击窗体中修改按钮转入宿舍修改窗体,运行效果如图 5-10 所示。注:该窗体的部分控件被其他模块调用,所以,在该窗体中将 TextBox 控件的 Modifiers 属性设为 Public。

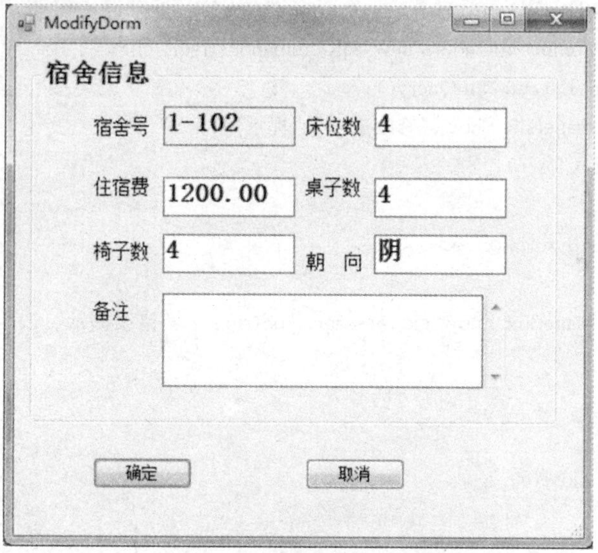

图 5-10 宿舍修改界面

```
public ModifyDorm()
{
    InitializeComponent();
}
```

```csharp
        SqlConnection cn = new SqlConnection(ConfigurationManager.ConnectionStrings["mycon"].ConnectionString.ToString());
        private void btnModify_Click(object sender, EventArgs e)
        {
            if (txtDormBed.Text.Trim() == "")
                MessageBox.Show("请输入床位数","提示");
            else
            {
                try
                {
                    cn.Open();
                    string strsql = string.Format("update dorm set dormBed='{0}',dormMoney={1},dormchair={2}, dormdesk={3}, dormdirection='{4}', dormremark='{5}' where dormID='{6}'", txtDormBed.Text, txtDormMoney.Text, txtDormChair.Text, txtDormDesk.Text, txtDormDirection.Text, txtdormremark.Text, txtDormID.Text);
                    SqlCommand cmd = new SqlCommand(strsql, cn);
                    cmd.ExecuteNonQuery();
                    MessageBox.Show("修改成功","提示");
                    this.Close();
                }
                catch (SqlException ex)
                {
                    MessageBox.Show(ex.Message.ToString(),"错误提示");
                }
                finally
                {
                    cn.Close();
                }
            }
        }
```

5.5.6 学生入住

5.5.6.1 学生入住安排

在主界面中选择"学生入住"/"学生入住安排"命令菜单,即可进入学生入住安排界面,如图5-11所示。

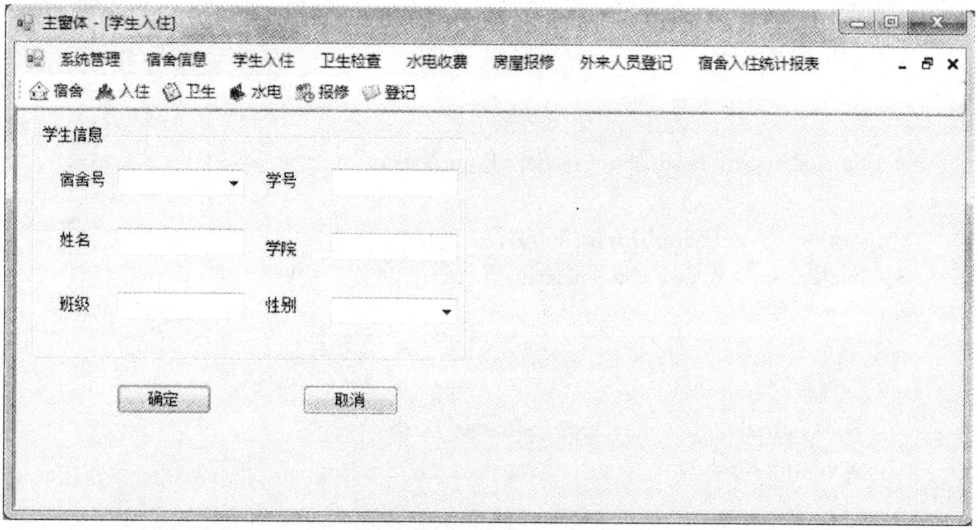

图 5-11 学生入住界面

该窗体中各个控件的名称、属性、事件、备注信息如表 5-14 所示。

表 5-14 学生入住界面控件设计

控件类型	控件名称	属性值	事件	备注
Form	addStudent	Text="学生入住"		
TextBox	txtsno			学生学号
	txtsname			学生姓名
	txtsdept			学生学院
	txtclass			学生班级
ComboBox	txtDormID			
	txtssex	Items:男、女		
Button	btnaddstudent	Text=确定	Click	
	btncancel	Text=取消	Click	

学生入住界面主要代码：

```
SqlConnection cn = new SqlConnection(ConfigurationManager. ConnectionStrings ["mycon"]. ConnectionString. ToString());
    public AddStudent()
```

```csharp
{
    InitializeComponent();
}

private void AddStudent_Load(object sender, EventArgs e)
{
    string str = "select dormID from dorm";
    SqlCommand cmd = new SqlCommand(str, cn);
    try
    {
        cn.Open();
        SqlDataReader dr = cmd.ExecuteReader();
        while (dr.Read())
        {
            txtDormID.Items.Add(dr[0].ToString());
        }
    }
    catch (SqlException ex)
    {
        MessageBox.Show(ex.Message.ToString());
    }
    finally
    {
        cn.Close();
    }
}

private void addstudent_Click(object sender, EventArgs e)
{
    string sql;
    //安排住宿时先判断该宿舍是否已经住满
    sql = "select dormBed-(select count(*) from student where dormID='" + txtDormID.Text.Trim() + "') from dorm where dormID='" + txtDormID.Text.Trim() + "'";
    SqlCommand cmd = new SqlCommand();
    cmd.CommandText = sql;
    cmd.Connection = cn;
```

```
            cn.Open();
            if (Convert.ToInt16(cmd.ExecuteScalar().ToString().Trim()) <= 0)
                MessageBox.Show("该房间已满","提示");
            else
            {
                sql = "select * from student where sno='" + txtsno.Text.Trim() + "'";
                cmd.CommandText = sql;
                if (cmd.ExecuteScalar() != null)
                    MessageBox.Show("学号重复","提示");
                else
                {
                    sql = string.Format("insert into student(sno,sname,ssex,sdept,sclass,dormID) values('{0}','{1}','{2}','{3}','{4}','{5}')", txtsno.Text, txtsname.Text, txtssex.Text, txtsdept.Text, txtclass.Text, txtDormID.Text);
                    cmd.CommandText = sql;
                    cmd.ExecuteNonQuery();
                    MessageBox.Show("添加成功","提示");
                    txtclass.Text = "";
                    txtDormID.Text = "";
                    txtsdept.Text = "";
                    txtsname.Text = "";
                    txtsno.Text = "";
                    txtssex.Text = "";
                }
            }
            cn.Close();
        }

        private void btncancel_Click(object sender, EventArgs e)
        {
            this.Close();
        }
```

5.5.6.2 学生查询

在主界面中选择"学生入住"/"学生查询"命令菜单,即可进入学生入住安排界面,如图5-12所示。

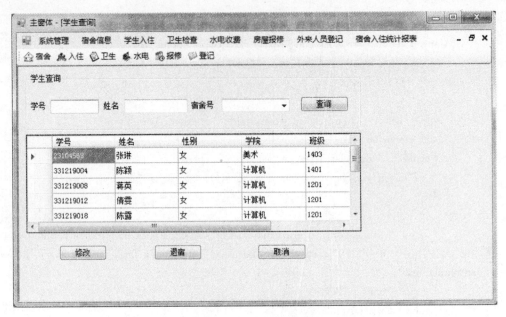

图 5-12 学生查询界面

该窗体中各个控件的名称、属性、事件、备注信息如表 5-15 所示。注：该窗体调用了 ModifyStudent 窗体中的部分控件，所以，在 ModifyStudent 窗体中将这些控件的 Modifiers 属性设为 Public。

表 5-15 学生查询界面控件设计

控件类型	控件名称	属性值	事件	备注
Form	searchStudent	Text="学生查询"		
TextBox	txtsno			学生学号
	txtsname			学生姓名
ComboBox	txtDormID			
DataGridView	dataGridView1			
Button	btnupdate	Text=修改	Click	
	btndel	Text=退宿	Click	
	btncancel	Text=取消	Click	

学生查询界面主要代码：

```csharp
    SqlConnection cn = new SqlConnection(ConfigurationManager.ConnectionStrings["mycon"].ConnectionString.ToString());
    DataSet ds=new DataSet();
    SqlDataAdapter adpter;
    void displaydb()
    {
        string str = "select sno 学号,sname 姓名,ssex 性别,sdept 学院,sclass 班级,dormID 宿舍号 from student";
        SqlCommand cmd = new SqlCommand(str, cn);
        try
        {
            adpter = new SqlDataAdapter(cmd);
            ds.Clear();
            adpter.Fill(ds, "s");
            dataGridView1.DataSource = ds.Tables["s"];
        }
        catch (SqlException ex)
        {
            MessageBox.Show(ex.Message.ToString());
        }
    }
    private void SearchStudent_Load(object sender, EventArgs e)
    {
        displaydb();

        string str = "select dormID from dorm";
        SqlCommand cmd = new SqlCommand(str, cn);
        cn.Open();
        SqlDataReader dr = cmd.ExecuteReader();
        while (dr.Read())
        {
            txtDormID.Items.Add(dr[0].ToString());
        }
        cn.Close();
    }
```

```csharp
private void btnsearch_Click(object sender, EventArgs e)
{
    string str = "select sno 学号,sname 姓名,ssex 性别,sdept 学院,sclass 班级,dormID 宿舍号 from student where 0=0";
    if (txtsno.Text != "")
        str += " and sno='" + txtsno.Text + "'";
    if (txtsname.Text != "")
        str += " and sname='" + txtsname.Text + "'";
    if (txtDormID.Text != "")
        str += " and dormID='" + txtDormID.Text + "'";
    SqlCommand cmd = new SqlCommand(str, cn);
    try
    {
        adpter = new SqlDataAdapter(cmd);
        ds.Clear();
        adpter.Fill(ds, "s");
        dataGridView1.DataSource = ds.Tables["s"];
    }
    catch (SqlException ex)
    {
        MessageBox.Show(ex.Message.ToString());
    }
    finally
    {
        txtDormID.Text = "";
        txtsname.Text = "";
        txtsno.Text = "";
    }
}

private void btnupdate_Click(object sender, EventArgs e)
{
    ModifyStudent modifystudent = new ModifyStudent();
    if (dataGridView1.DataSource != null && dataGridView1.CurrentCell.RowIndex >= 0 && dataGridView1.CurrentCell != null)
    {
```

```
                modifystudent.txtsno.Text = ds.Tables["s"].Rows[dataGridView1.CurrentCell.
RowIndex]["学号"].ToString().Trim();
                modifystudent.txtsname.Text = ds.Tables["s"].Rows[dataGridView1.CurrentCell.
RowIndex]["姓名"].ToString().Trim();
                modifystudent.txtssex.Text = ds.Tables["s"].Rows[dataGridView1.CurrentCell.
RowIndex]["性别"].ToString().Trim();
                modifystudent.txtsdept.Text = ds.Tables["s"].Rows[dataGridView1.CurrentCell.
RowIndex]["学院"].ToString().Trim();
                modifystudent.txtclass.Text = ds.Tables["s"].Rows[dataGridView1.CurrentCell.
RowIndex]["班级"].ToString().Trim();
                modifystudent.txtDormID.Text = ds.Tables["s"].Rows[dataGridView1.CurrentCell.
RowIndex]["宿舍号"].ToString().Trim();
            modifystudent.ShowDialog();
            displaydb();
        }
    }

    private void btndel_Click(object sender, EventArgs e)
    {
        SqlCommandBuilder cmdbuilder = new SqlCommandBuilder(adpter);
        if (MessageBox.Show("真的删除吗?", "提示", MessageBoxButtons.YesNoCancel, MessageBoxIcon.Warning) == DialogResult.Yes)
        {
            DataRow delrow = ds.Tables["s"].Rows[dataGridView1.CurrentCell.RowIndex];
            delrow.Delete();
            adpter.Update(ds, "s");
            MessageBox.Show("删除成功");
        }

        string str = "select sno 学号,sname 姓名,ssex 性别,sdept 学院,sclass 班级,dormID 宿舍号 from student where 0=0";
        if (txtsno.Text != "")
            str += " and sno='" + txtsno.Text + "'";
        if (txtsname.Text != "")
            str += " and sname='" + txtsname.Text + "'";
        if (txtDormID.Text != "")
            str += " and dormID='" + txtDormID.Text + "'";
```

```csharp
        SqlCommand cmd = new SqlCommand(str, cn);
        try
        {
            adpter = new SqlDataAdapter(cmd);
            ds.Clear();
            adpter.Fill(ds, "s");
            dataGridView1.DataSource = ds.Tables["s"];
        }
        catch (SqlException ex)
        {
            MessageBox.Show(ex.Message.ToString());
        }
    }

    private void btncancel_Click(object sender, EventArgs e)
    {
        this.Close();
    }
```

5.5.6.3 学生修改

在主界面中选择"学生入住"/"学生查询"命令菜单,即可进入学生入住安排界面,如图 5-12 所示,点击修改按钮转入学生修改窗体,运行效果如图 5-13 所示。注:该窗体的部分控件被其他模块调用,所以,在该窗体中将 TextBox、ComboBox 控件的 Modifiers 属性设为 Public。

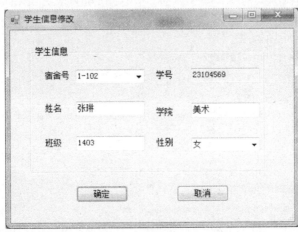

图 5-13 修改学生信息界面

修改学生信息代码：

```csharp
public ModifyStudent()
{
    InitializeComponent();
}
    SqlConnection cn = new SqlConnection(ConfigurationManager.ConnectionStrings["mycon"].ConnectionString.ToString());
    private void btnmodify_Click(object sender, EventArgs e)
    {
        if (txtsname.Text == "" || txtDormID.Text == "" || txtsno.Text == "")
            MessageBox.Show("请输入完整信息");
        else
        {
            cn.Open();
            string sql = "select * from dorm where dormID='" + txtDormID.Text.Trim() + "'";
            SqlCommand cmd = new SqlCommand(sql, cn);
            if (cmd.ExecuteScalar() == null)
            {
                MessageBox.Show("没有该宿舍号", "提示");
            }
            else
            {
                sql = string.Format("update student set dormID='{0}',ssex='{1}',sname='{2}',sdept='{3}',sclass='{4}' where sno='{5}' ", txtDormID.Text, txtssex.Text, txtsname.Text, txtsdept.Text, txtclass.Text, txtsno.Text);
                cmd.CommandText = sql;
                cmd.Connection = cn;
                cmd.ExecuteNonQuery();
                MessageBox.Show("修改成功", "提示");
                this.Close();
            }
            cn.Close();
        }
    }

    private void btncancel_Click(object sender, EventArgs e)
```

```
{
    this.Close();
}

private void ModifyStudent_Load(object sender, EventArgs e)
{
    //combobox中只显示有剩余床位的宿舍号,通过视图实现
    string sql = "select * from numberofbed";
    SqlCommand cmd1 = new SqlCommand();
    cmd1.CommandText = sql;
    cmd1.Connection = cn;
    cn.Open();
    SqlDataReader dr = cmd1.ExecuteReader();
    while(dr.Read())
    {
        if(Convert.ToInt16(dr[1])!=0)
            txtDormID.Items.Add(dr[0].ToString());
    }
    cn.Close();
}
```

5.5.7 卫生检查

在主界面中选择"卫生检查"/"添加成绩"/"查询成绩"命令菜单,即可进入卫生成绩添加、卫生成绩查询界面,运行效果分别如图5-14、图5-15所示,点击查询窗口的修改,进入修改窗体,运行效果如图5-16所示。实现过程与上述其他窗体实现过程雷同,这里不再赘述。

5.5.7.1 添加成绩

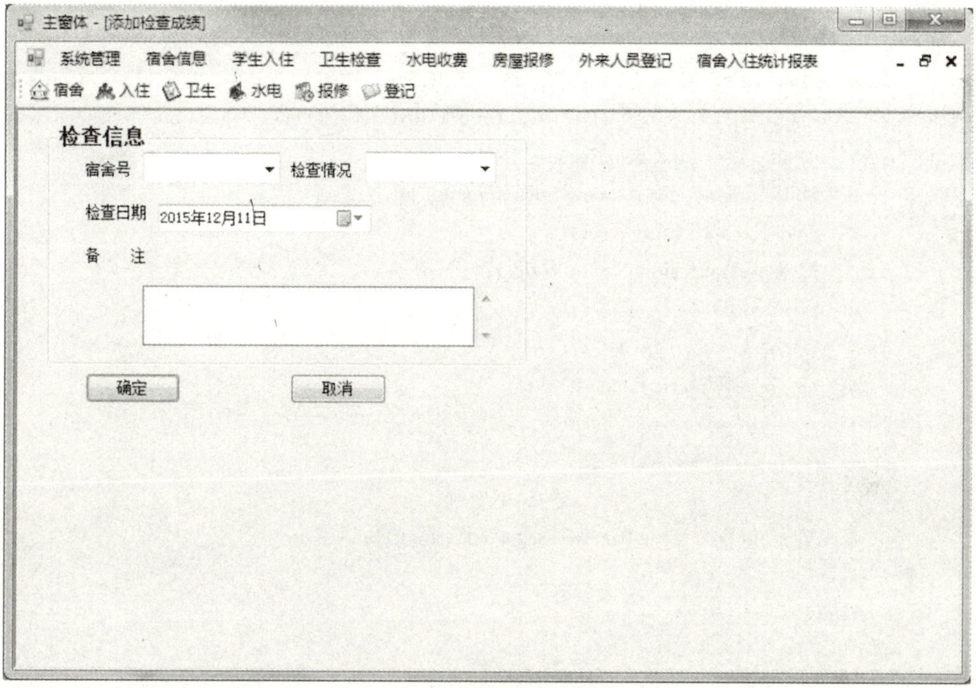

图 5-14 添加检查成绩界面

添加成绩代码：

```
public AddCheck()
{
    InitializeComponent();
}
   SqlConnection cn = new SqlConnection(ConfigurationManager.ConnectionStrings["mycon"].ConnectionString.ToString());
   DataSet ds = new DataSet();
   //SqlDataAdapter adpter;
   private void btnaddcheck_Click(object sender, EventArgs e)
   {
       if (txtDormID.Text == "")
           MessageBox.Show("请选择宿舍号");
       else
       {
```

```csharp
            try
            {
                cn.Open();
                string sql = string.Format("insert into checkinformation(dormID, CheckDate, CheckState, CheckRemark) values('{0}','{1}','{2}','{3}')", txtDormID.Text, checkTime.Value.Date.ToString("yyyy-MM-dd"), txtcheckstate.Text, txtCheckRemark.Text);
                SqlCommand cmd = new SqlCommand(sql, cn);
                cmd.ExecuteNonQuery();
                MessageBox.Show("插入成功");
                txtDormID.Text = "";
                txtcheckstate.Text = "";
                txtCheckRemark.Text = "";
            }
            catch (SqlException ex)
            {
                MessageBox.Show(ex.Message.ToString());
            }
            finally
            {
                cn.Close();
            }
        }

        private void AddCheck_Load(object sender, EventArgs e)
        {
            string str = "select dormID from dorm";
            SqlCommand cmd = new SqlCommand(str, cn);

            try
            {
                cn.Open();
                SqlDataReader dr = cmd.ExecuteReader();
                while (dr.Read())
                {
                    txtDormID.Items.Add(dr[0].ToString());
                }
```

```
    }
    catch (SqlException ex)
    {
        MessageBox.Show(ex.Message.ToString());
    }
    finally
    {
        cn.Close();
    }
}

private void btncancel_Click(object sender, EventArgs e)
{
    this.Close();
}
```

5.5.7.2 查询成绩

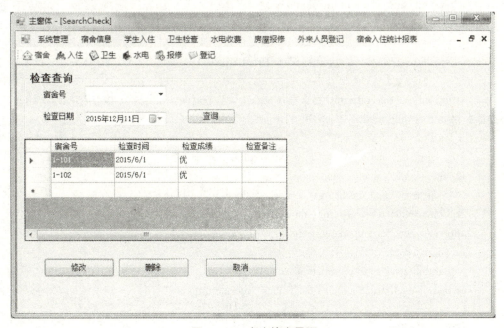

图 5-15 查询检查界面

查询成绩代码:

```csharp
public SearchCheck()
{
    InitializeComponent();
}
    SqlConnection cn = new SqlConnection(ConfigurationManager.ConnectionStrings["mycon"].ConnectionString.ToString());
    DataSet ds = new DataSet();
    SqlDataAdapter adpter;
    void displaydb()
    {
        string sql = "select dormID 宿舍号,CheckDate 检查时间,CheckState 检查成绩,CheckRemark 检查备注 from checkinformation";
        SqlCommand cmd = new SqlCommand(sql, cn);
        adpter = new SqlDataAdapter(cmd);
        ds.Clear();
        adpter.Fill(ds, "checkinformation");
        dataGridView1.DataSource = ds.Tables["checkinformation"];
    }
    private void btnsearch_Click(object sender, EventArgs e)
    {
        string sql = "select dormID 宿舍号,CheckDate 检查时间,CheckState 检查成绩,CheckRemark 检查备注 from checkinformation where 0=0";
        if (txtDormID.Text != "")
            sql += " and dormID='" + txtDormID.Text + "'";
        if (checkTime.Value != null)
            sql += " and checkdate='" + checkTime.Value.Date + "'";
        SqlCommand cmd = new SqlCommand(sql, cn);
        adpter = new SqlDataAdapter(cmd);
        ds.Clear();
        adpter.Fill(ds, "checkinformation");
        dataGridView1.DataSource = ds.Tables["checkinformation"];

        txtDormID.Text = "";
    }
```

```csharp
private void SearchCheck_Load(object sender, EventArgs e)
{
    displaydb();

    string str = "select dormID from dorm";
    SqlCommand cmd = new SqlCommand(str, cn);
    cn.Open();
    SqlDataReader reader=cmd.ExecuteReader();
    while (reader.Read())
    {
        txtDormID.Items.Add(reader[0].ToString());
    }
    cn.Close();
}
Modifycheck modifycheck = new Modifycheck();
private void btnupdate_Click(object sender, EventArgs e)
{
    if (dataGridView1.DataSource != null && dataGridView1.CurrentCell.RowIndex >= 0 && dataGridView1.CurrentCell != null)
    {
        modifycheck = new Modifycheck();
        modifycheck.txtDormID.Text = ds.Tables["checkinformation"].Rows[dataGridView1.CurrentCell.RowIndex][0].ToString().Trim();
        modifycheck.checkTime.Text = ds.Tables[0].Rows[dataGridView1.CurrentCell.RowIndex][1].ToString().Trim();
        modifycheck.txtCheckRemark.Text = ds.Tables[0].Rows[dataGridView1.CurrentCell.RowIndex]["CheckRemark"].ToString().Trim();
        modifycheck.txtcheckstate.Text = ds.Tables[0].Rows[dataGridView1.CurrentCell.RowIndex]["checkstate"].ToString().Trim();
        modifycheck.ShowDialog();
    }
    displaydb();
}
```

5.5.7.3 修改成绩

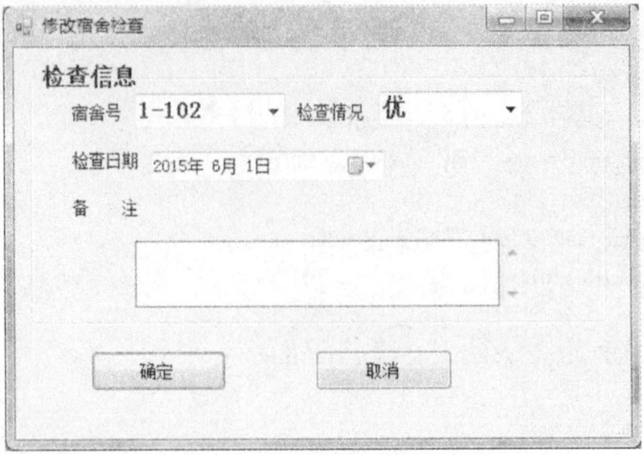

图 5-16 卫生成绩修改界面

修改成绩代码:

```
    SqlConnection cn = new SqlConnection(ConfigurationManager.ConnectionStrings["mycon"].ConnectionString.ToString());
    public Modifycheck()
    {
        InitializeComponent();
    }

    private void btnmodifycheck_Click(object sender, EventArgs e)
    {
        if (txtDormID.Text.Trim() == "")
            MessageBox.Show("请填写宿舍号");
        else
        {
            string str = string.Format("update checkinformation set CheckDate='{0}', CheckState='{1}', CheckRemark='{2}' where dormID='{3}'", checkTime.Value.ToString("yyyy-MM-dd"), txtcheckstate.Text, txtCheckRemark.Text, txtDormID.Text);
            SqlCommand cmd = new SqlCommand(str, cn);
            cn.Open();
            cmd.ExecuteNonQuery();
            MessageBox.Show("修改成功", "提示");
```

```
            cn.Close();
            this.Close();
        }
    }

    private void btncancel_Click(object sender, EventArgs e)
    {
        this.Close();
    }
```

5.5.8 水电收费

5.5.8.1 水电抄表

在主界面中选择"水电收费"/"水电抄表"命令菜单,即可进入水电抄表界面,如图5-17所示,控件相关属性请读者自行设置。

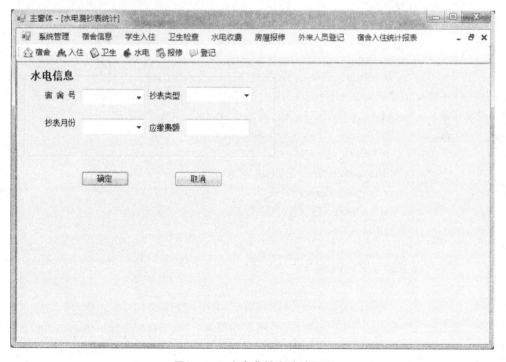

图 5-17 水电费抄表统计界面

水电费抄表统计代码：

```csharp
public partial class AddCheck : Form
{
    public AddCheck()
    {
        InitializeComponent();
    }
    SqlConnection cn = new SqlConnection(ConfigurationManager.ConnectionStrings["mycon"].ConnectionString.ToString());
    DataSet ds = new DataSet();
    //SqlDataAdapter adpter;
    private void btnaddcheck_Click(object sender, EventArgs e)
    {
        if (txtDormID.Text == "")
            MessageBox.Show("请选择宿舍号");
        else
        {
            try
            {
                cn.Open();
                string sql = string.Format("insert into checkinformation(dormID,CheckDate,CheckState,CheckRemark) values('{0}','{1}','{2}','{3}')", txtDormID.Text, checkTime.Value.Date.ToString("yyyy-MM-dd"), txtcheckstate.Text, txtCheckRemark.Text);
                SqlCommand cmd = new SqlCommand(sql, cn);
                cmd.ExecuteNonQuery();
                MessageBox.Show("插入成功");
                txtDormID.Text = "";
                txtcheckstate.Text = "";
                txtCheckRemark.Text = "";
            }
            catch (SqlException ex)
            {
                MessageBox.Show(ex.Message.ToString());
            }
            finally
            {
```

```
            cn.Close();
        }
    }
}

private void AddCheck_Load(object sender, EventArgs e)
{
    string str = "select dormID from dorm";
    SqlCommand cmd = new SqlCommand(str, cn);

    try
    {
        cn.Open();
        SqlDataReader dr = cmd.ExecuteReader();
        while (dr.Read())
        {
            txtDormID.Items.Add(dr[0].ToString());
        }
    }
    catch (SqlException ex)
    {
        MessageBox.Show(ex.Message.ToString());
    }
    finally
    {
        cn.Close();
    }
}

private void btncancel_Click(object sender, EventArgs e)
{
    this.Close();
}
```

5.5.8.2 水电收缴

在主界面中选择"水电收费"/"水电收缴"命令菜单,即可进入水电费收缴界面,如图5-18所示,控件相关属性请读者自行设置。

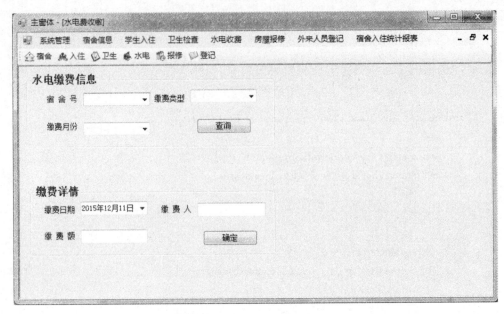

图 5-18 水电费收缴、查询界面

水电费收缴、查询代码：

```
    SqlConnection cn = new SqlConnection(ConfigurationManager.ConnectionStrings["mycon"].ConnectionString.ToString());
    private void btnsearchcharge_Click(object sender, EventArgs e)
    {
        if (txtdormID.Text == "" || txtchargemonth.Text == "" || txtchargetype.Text == "")
            MessageBox.Show("请选择完整信息","提示");
        else
        {
            cn.Open();
            string str = string.Format("select chargemoney,chargeperson,chargedate,flag from charge where dormID='{0}' and chargetype='{1}'and chargemonth='{2}'", txtdormID.Text, txtchargetype.Text, txtchargemonth.Text);
            SqlCommand cmd = new SqlCommand(str, cn);
            SqlDataReader dr = cmd.ExecuteReader();
            try
            {
                if (dr.Read())
                {
```

第 5 章 课程设计案例

```
                if (Convert.ToBoolean(dr[3]) == true)
                {
                    label7.Text = "该宿舍已经缴费成功";
                    txtchargemoney.Text = dr[0].ToString();
                    txtchargeperson.Text = dr[1].ToString();
                    dateTimePicker1.Text = Convert.ToDateTime(dr[2]).ToString("yyyy-MM-dd");
                }
                else
                {
                    label7.Text = "该宿舍尚未缴费,请缴费";
                    txtchargemoney.Text = dr[0].ToString();
                }
            }
            else
            {
                label7.Text = "尚未抄表,无法缴费";
            }
        }
        catch (SqlException ex)
        {
            MessageBox.Show(ex.Message.ToString());
        }
        finally
        {
            cn.Close();
        }
    }
}

private void AddCharge_Load(object sender, EventArgs e)
{
    string str = "select dormID from dorm";
    SqlCommand cmd = new SqlCommand(str, cn);
    try
    {
```

```csharp
            cn.Open();
            SqlDataReader dr = cmd.ExecuteReader();
            while (dr.Read())
            {
                txtdormID.Items.Add(dr[0].ToString());
            }
        }
        catch (SqlException ex)
        {
            MessageBox.Show(ex.Message.ToString());
        }
        finally
        {
            cn.Close();
        }
    }

    private void btnaddcharge_Click(object sender, EventArgs e)
    {
        SqlCommand cmd = new SqlCommand();
        if (txtchargeperson.Text == "" || txtchargemoney.Text == "")
            MessageBox.Show("请完整选择信息");
        else
        {
            cn.Open();
            string sql = string.Format("update charge set chargemoney={0},chargeperson='{1}',chargedate='{2}',flag='true' where dormID='{3}' and chargetype='{4}' and chargemonth='{5}'", txtchargemoney.Text, txtchargeperson.Text, dateTimePicker1.Value.ToString("yyyy-MM-dd"), txtdormID.Text, txtchargetype.Text, txtchargemonth.Text);
            cmd.CommandText = sql;
            cmd.Connection = cn;
            cmd.ExecuteNonQuery();
            MessageBox.Show("缴费成功");
            txtchargemoney.Text = "";
            txtchargemonth.Text = "";
            txtchargeperson.Text = "";
```

```
            txtchargetype.Text = "";
            txtdormID.Text = "";
            cn.Close();
        }
```

5.5.9 房屋报修

报修登记、报修查询运行效果分别如下图 5-19、5-20 所示,实现方法与上述其他模块雷同,这里不再赘述,请读者参考代码自行完成。

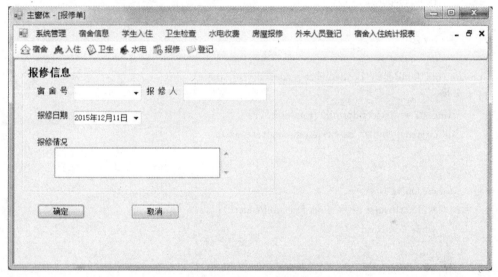

图 5-19 报修单填写界面

报修并填写代码:

```
    SqlConnection cn = new SqlConnection(ConfigurationManager.ConnectionStrings ["mycon"].ConnectionString.ToString());
    private void btnaddrepair_Click(object sender, EventArgs e)
    {
        if (txtDormID.Text == "" || txtRepairRemark.Text == "")
            MessageBox.Show("宿舍号或报修情况不能为空");
        else
        {
```

```csharp
            string sql = string.Format("insert into repair(dormID, repairperson, addrepairdate, repairremark) values ('{0}','{1}','{2}','{3}')", txtDormID.Text, txtreportperson.Text, dateTimePicker2.Value.ToString("yyyy-MM-dd"), txtRepairRemark.Text);
            SqlCommand cmd = new SqlCommand(sql, cn);
            cn.Open();
            cmd.ExecuteNonQuery();
            MessageBox.Show("添加成功");
            txtDormID.Text = "";
            txtRepairRemark.Text = "";
            txtreportperson.Text = "";
        }
    }

    private void AddRepair_Load(object sender, EventArgs e)
    {
        string str = "select dormID from dorm";
        SqlCommand cmd = new SqlCommand(str, cn);
        try
        {
            cn.Open();
            SqlDataReader dr = cmd.ExecuteReader();
            while (dr.Read())
            {
                txtDormID.Items.Add(dr[0].ToString());
            }
        }
        catch (SqlException ex)
        {
            MessageBox.Show(ex.Message.ToString());
        }
        finally
        {
            cn.Close();
        }
    }
```

第 5 章　课程设计案例

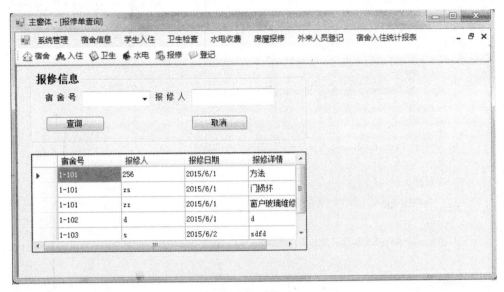

图 5-20　报修单查询界面

报修并查询代码：

```
public SearchRepair()
{
    InitializeComponent();
}
    SqlConnection cn = new SqlConnection(ConfigurationManager.ConnectionStrings["mycon"].ConnectionString.ToString());
    DataSet ds = new DataSet();
    SqlDataAdapter adpter;

    private void SearchRepair_Load(object sender, EventArgs e)
    {
        string sql = "select dormID 宿舍号, repairperson 报修人, addrepairdate 报修日期, repairremark 报修详情 from rcpair";
        adpter = new SqlDataAdapter(sql, cn);
        adpter.Fill(ds, "repair");
        dataGridView1.DataSource = ds.Tables["repair"];

        string str = "select dormID from dorm";
```

```csharp
        SqlCommand cmd = new SqlCommand(str, cn);
        try
        {
            cn.Open();
            SqlDataReader dr = cmd.ExecuteReader();
            while (dr.Read())
            {
                txtDormID.Items.Add(dr[0].ToString());
            }
        }
        catch (SqlException ex)
        {
            MessageBox.Show(ex.Message.ToString());
        }
        finally
        {
            cn.Close();
        }
    }

    private void btnsearchrepair_Click(object sender, EventArgs e)
    {
        string sql = string.Format("select dormID 宿舍号, repairperson 报修人, addrepairdate 报修日期, repairremark 报修详情 from repair where 0=0");
        if (txtDormID.Text != "")
            sql += " and dormID='" + txtDormID.Text + "'";
        if (txtreportperson.Text != "")
            sql += " and reportperson='" + txtreportperson.Text + "'";
        SqlCommand cmd = new SqlCommand(sql, cn);
        adpter = new SqlDataAdapter(cmd);
        ds.Clear();
        adpter.Fill(ds, "report");
        dataGridView1.DataSource = ds.Tables["report"];
    }
```

5.5.10 外来人员登记

来访登记、来访查询运行效果分别如下图 5-21、5-22 所示,实现方法与上述其他模块雷同,这里不再赘述,请读者参考代码自行完成。

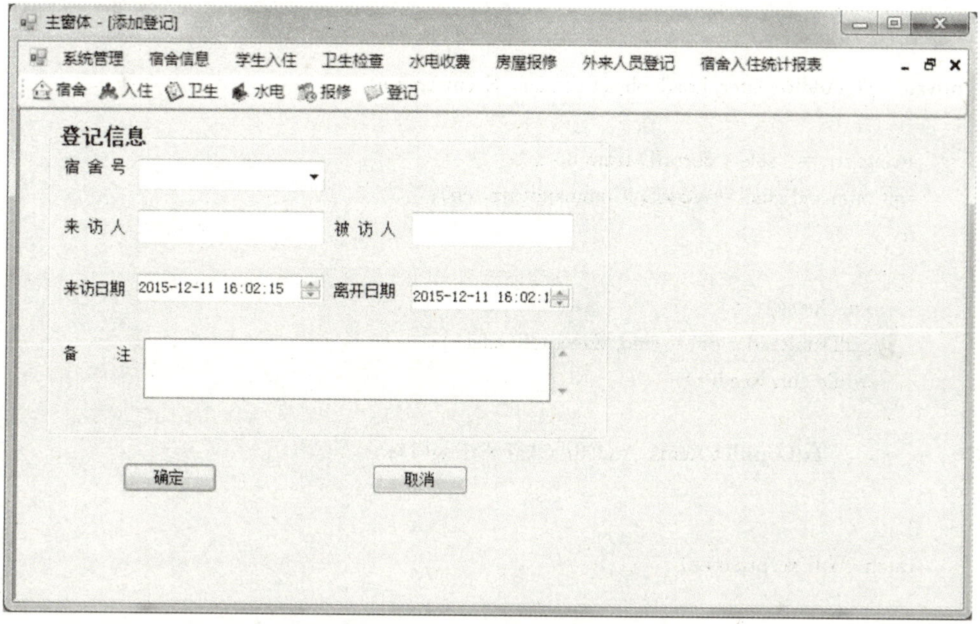

图 5-21 来访登记界面

来访登记代码:

```
    SqlConnection cn = new SqlConnection(ConfigurationManager.ConnectionStrings["mycon"].ConnectionString.ToString());
    private void btnAddregister_Click(object sender, EventArgs e)
    {
        string sql = string.Format(" insert into register(dormID, vistor, bevisted, vistdatetime, leaverdatetime, registerremark) values('{0}','{1}','{2}','{3}','{4}','{5}')", txtDormID.Text, txtVistor.Text, txtBevisted.Text, ArriveTime.Value.ToString("yyyy-MM-dd hh:mm:ss"), LeaveTime.Value.ToString("yyyy-MM-dd hh:mm:ss"), txtregsterRemark.Text);
        SqlCommand cmd = new SqlCommand(sql, cn);
        cn.Open();
        cmd.ExecuteNonQuery();
        MessageBox.Show("登记成功");
```

```csharp
            txtDormID.Text = "";
            txtBevisted.Text = "";
            txtVistor.Text = "";
            txtregsterRemark.Text = "";
            this.Close();
        }

        private void AddRegister_Load(object sender, EventArgs e)
        {
            string str = "select dormID from dorm";
            SqlCommand cmd = new SqlCommand(str, cn);
            try
            {
                cn.Open();
                SqlDataReader dr = cmd.ExecuteReader();
                while (dr.Read())
                {
                    txtDormID.Items.Add(dr[0].ToString());
                }
            }
            catch (SqlException ex)
            {
                MessageBox.Show(ex.Message.ToString());
            }
            finally
            {
                cn.Close();
            }

            ArriveTime.CustomFormat = "yyyy-MM-dd HH:mm:ss";
            this.ArriveTime.Format = DateTimePickerFormat.Custom;
            this.ArriveTime.ShowUpDown = true;
            LeaveTime.CustomFormat = "yyyy-MM-dd HH:mm:ss";
            this.LeaveTime.Format = DateTimePickerFormat.Custom;
            this.LeaveTime.ShowUpDown = true;
        }
```

第5章 课程设计案例

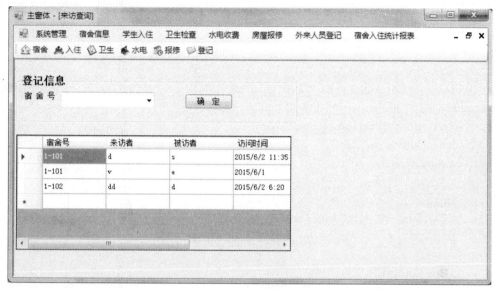

图5-22 来访查询界面

来访查询代码：

```
public SearchRegister()
{
    InitializeComponent();
}
SqlConnection cn = new SqlConnection(ConfigurationManager.ConnectionStrings["mycon"].ConnectionString.ToString());
SqlDataAdapter adpter;
DataSet ds=new DataSet();
private void btnAddregister_Click(object sender, EventArgs e)
{
    string sql = "select dormID 宿舍号,vistor 来访者,bevisted 被访者,vistdatetime 访问时间,leaverdatetime 离开时间,registerremark 备注 from register where 0=0";
    if (txtDormID.Text != "")
        sql += " and dormID='" + txtDormID.Text + "'";
    adpter = new SqlDataAdapter(sql, cn);
    ds.Clear();
    adpter.Fill(ds, "register");
    dataGridView1.DataSource = ds.Tables["register"];
```

}

```
private void SearchRegister_Load(object sender, EventArgs e)
{
    string sql = "select dormID 宿舍号,vistor 来访者,bevisted 被访者,vistdatetime 访问时间,leaverdatetime 离开时间,registerremark 备注 from register";
    adpter = new SqlDataAdapter(sql, cn);
    adpter.Fill(ds, "register");
    dataGridView1.DataSource = ds.Tables["register"];

    string str = "select dormID from dorm";
    SqlCommand cmd = new SqlCommand(str, cn);
    try
    {
        cn.Open();
        SqlDataReader dr = cmd.ExecuteReader();
        while (dr.Read())
        {
            txtDormID.Items.Add(dr[0].ToString());
        }
    }
    catch (SqlException ex)
    {
        MessageBox.Show(ex.Message.ToString());
    }
    finally
    {
        cn.Close();
    }
}
```

5.5.11 宿舍入住统计报表

宿舍入住统计报表可通过视图统计各宿舍已经入住的学生人数及剩余的床位数,使用报表实现,运行效果如图 5-23 所示。

实现过程如下:

1. 在 SQL Server 2008 R2 中启动 SQL Server Business Intelligence Development

Studio,新建报表项目1如图5-24所示,新建数据源如图5-25所示,新建数据集如图5-26所示,新建报表如图5-27所示。

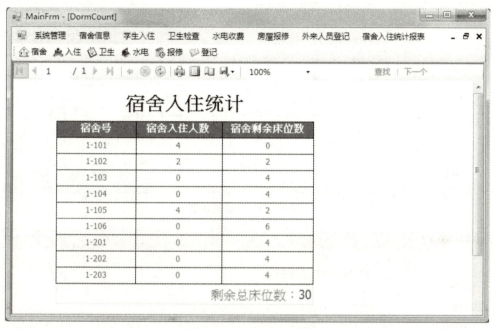

图5-23 报表统计界面

图5-24 新建报表项目1

图 5-25 新建数据源

图 5-26 新建数据集

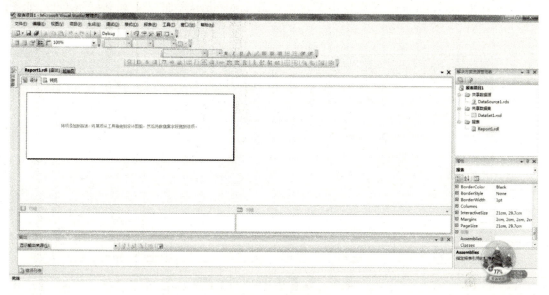

图 5-27 新建报表

2. 右击插入文本框,输入标题"宿舍入住统计"设置字体字号;插入表格,设置表格内容,过程如图 5-28 所示。

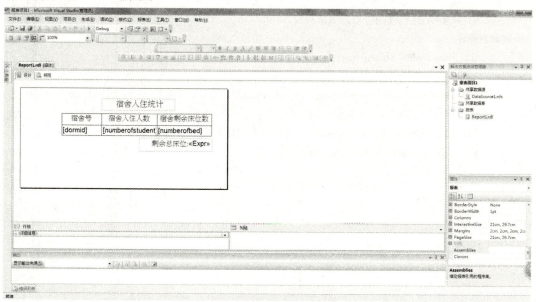

图 5-28 报表设计过程

3. 在 DormMangement 项目中新建窗体 DormCount，添加控件 reportViewer1，如图 5-29 所示。

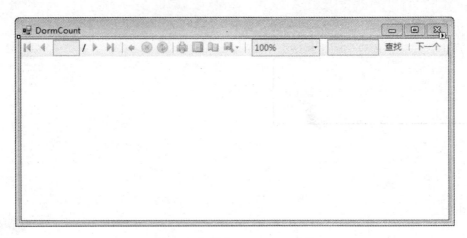

图 5-29 新建 DormCount 窗体

报表显示代码：

```
private void DormCount_Load(object sender, EventArgs e)
{
    // TODO: 这行代码将数据加载到表"DormMangementDataSet.dormcount"中。您可以根据需要移动或删除它。
    this.dormcountTableAdapter.Fill(this.DormMangementDataSet.dormcount);

    this.reportViewer1.ServerReport.ReportServerUrl = new Uri("http://202.195.126.161/ReportServer");
    this.reportViewer1.ServerReport.ReportPath = "/报表项目1/Report1";
    this.reportViewer1.RefreshReport();
}
```

第 6 章　数据库课程设计选题

数据库课程设计是学生实践课程理论知识,真正掌握数据库应用系统开发技能,直至最终能解决复杂多样的实际数据管理问题的重要教学环节。数据库原理及应用课程是一门实践性较强的专业基础理论课,通过课程设计能有效地加深对数据库基础理论和基本知识的理解,掌握开发一个小型数据库系统时设计数据库结构的基本方法,提高运用数据库解决实际问题的能力。

本章列举了几个课程设计题目及要求,供学生在课程设计中参考。数据库的应用非常广泛,学生可在教师指导下根据实际需求自己选择设计题目。

6.1　课程设计基本要求

课程设计能训练学生灵活应用所学课程知识的能力,能够综合运用所学的理论知识和方法独立分析和解决问题;初步掌握软件开发过程的问题分析、系统设计、程序编码、测试等基本方法和技能;训练用系统的观点和软件开发一般规范进行软件开发,巩固、深化学生的理论知识,提高编程水平;培养学生调查研究,查阅技术文献、资料、手册以及编写技术文献的能力;并在此过程中培养他们严谨的科学态度和良好的工作作风。

课程设计要求:

(1) 理解和掌握 E-R 图的设计方法和关系模式的转换方法。根据课题的要求列出实体,联系画出 E-R 图并注明主码和外码;

(2) 创建数据库及各种数据库对象(至少创建三个以上的数据表),根据题目的要求编写存储过程、触发器等;

(3) 选用熟悉的主流软件平台作为开发工具,完成系统的开发。系统基本功能应有数据录入、修改、删除、数据完整性检验、查询(分精确查询与模糊查询)、报表打印等;

(4) 通过调查研究或运用 Internet,收集和调查有关资料;基本掌握撰写设计报告的基本步骤和写作方法。最终提交完整的设计报告。

6.2 课程设计选题

6.2.1 工资管理系统

工资管理系统是企、事业单位常见的计算机信息管理系统。它的主要任务是对各种工资信息进行日常的管理,如工资信息的输入、查询、修改、增加、删除,迅速准确地完成各种工资信息的统计计算和汇总,快速打印出报表。在建立与实现工资管理系统时,应进行功能划分,能够由数据库管理系统完成的功能应尽可能由数据库管理系统守成,这样才能够充分发挥数据库管理系统高效、完全、可靠、便捷的性能,减少编程人员的工作量。

工资管理系统是数据库管理系统的一个比较典型的应用,它具有大多数数据库应用系统的特征,包含了教材中所涉及的大多数 SQL Server 数据库对象。在进行课程设计时,应尽可能使用 SQL Server 的功能完成下列功能设计的各项操作。

6.2.1.1 系统功能设计

(1) 信息输入功能

① 输入员工的基本信息。包括:员工编号、姓名、性别、出生年月、参加工作时间、所属部门、职务、职称、政治面貌、婚姻状况等基本信息。

② 输入员工的工资信息。包括:基本工资、岗位工资、住房补贴、津贴、工会会费、水电费、住房公积金、养老保险、奖惩。

③ 输入员工的部门信息。包括:部门编号、部门名称、部门负责人、部门人数。

(2) 数据修改删除功能

① 修改和删除员工的基本信息。当单位人员的信息发生变化,如职称的改变,工作部门变动,或调离本单位等,系统应能修改员工的信息或将其从员工信息表中删除。

② 修改和删除员工的工资信息。员工升职加薪、工资普调是企业中常见的事情,这就需要系统能方便对员工工资进行个别及批量的修改;如员工调离本单位,就应当从员工信息表中删除这个员工信息,将其工资信息表中相应的数据删除。

③ 修改和删除部门信息。当撤消一个部门时,能从部门信息表中将其删除。而当一个部门的名称发生改变时,系统中所有该项部门的名称应一致的改变。

(3) 查询和统计功能

① 能够查询指定员工的基本信息。

② 能够查询指定某个部门或全部部门的基本信息。

③ 查询某个员工的工资信息。

④ 统计、汇总指定年份企业各部门或某个部门每个月或全年工资总额,汇总各部门的

人数或本单位的总人数。

⑤ 工资表月工资记录的生成功能。生成当月所有员工或某个部门的工资记录,同时能进行员工工资的计算,即计算应发金额、应扣金额及实发金额。

6.2.1.2 数据完整性设计

为了保证数据库系统的正确性、完备性和一致性,就必须进行数据完整性设计。就本设计而言应考虑实施如下数据完整性:

(1) 给每个表实施主键及外键约束。

(2) 设定缺省约束。如员工性别。

(3) 设置非空约束。如员工姓名。

(4) 实施 CHECK 约束,如养老保险的金额大于 0。

(5) 实施规则,如政治面貌必须是"党员"、"团员"及"群众"三个选项之一。

6.2.1.3 数据库对象的设计

为充分发挥数据库的效能,保证数据库的安全性,提高数据库管理系统的执行效率,可以考虑使用视图、存储过程及表的触发器来实现某些功能。本设计可考虑如下数据库对象:

(1) 指定员工或某个部门的信息查询。可以设计一个存储过程,以员工编号或部门编号为输入参数返回指定员工或部门的基本信息。

(2) 统计指定年份整个企业或各部门每个月份的工资总额。设计一个以年份为输入参数,以总工资金额为输出参数的存储过程,返回每个月份企业工资支出的总额。

(3) 浏览工资表。设计一个视图,返回当前月份所有员工或某个部门员工的工资信息。

(4) 为提高检索性能,为表创建索引。

(5) 为新调入/调出/内部调动人员创建 INSERT、DELETE、UPDATE 触发器,实现部门人数的自动更新。

6.2.2 教务管理系统

教务管理系统是学校常见的计算机信息管理系统。它的主要任务是对各种教学信息进行日常的管理,如课程管理、任课教师管理等。迅速准确地完成各种计算和汇总,快速打印出报表。充分发挥数据库管理系统高效、完全、可靠、便捷的性能,减少教务人员的工作量。

6.2.2.1 系统功能设计

负责教务工作的教师的日常工作包括:

(1) 每学期开始时打印每个班级的基本信息(班名、入学时间、班长等)及学生(学号、姓名、性别、出生日期、照片、政治面貌、联系方式——如宿舍号,电话,EMAIL 等)的基本状况报表。

(2) 每学期重新打印一份教师名单,包含教师的基本信息(工作证号码、姓名、性别、出生日期、职称、职务、政治面貌、办公室房间号,电话号码等)。

(3) 采用友好的界面对系、班级、学生、教师、课程、选课等内容进行增、删、改。

(4) 具有方便的查询功能,例如,对于学生,可以按照学生的学号、姓名、年龄、性别、系别等属性的任意组合条件进行查询。同样为课程以及选课等内容进行查询。

(5) 具有丰富的报表统计功能,例如,对于学生选课信息,可以进行如下报表汇总操作:

① 打印出某学生某学期所选修的全部课程的学分、学时以及成绩。

② 打印出某教师所讲授的全部课程的信息。

③ 打印出某班某学期所有学生按总成绩降序列出的学号、姓名以及总绩报表。

④ 打印出各系具有的各级职称的教师人数。

⑤ 打印出该学院所开设的各门课程的名称、学时以及必修课的名称和学时。

6.2.2.2 数据库完整性设计

为了保证数据库系统的正确性、完备性和一致性,就必须进行数据完整性设计。就本设计而言应考虑实施如下数据完整性:

(1) 给每个表实施主键及外键约束。

(2) 设定缺省约束。如教师、学生的性别。

(3) 设置非空约束。如教师、学生的姓名。

(4) 实施 CHECK 约束。如学生入学时间在一定范围。

(5) 实施规则。如政治面貌必须是"党员"、"团员"及"群众"三个选项之一。

6.2.2.3 SQL Server 数据库对象设计

为充分发挥数据库的效能,保证数据库的安全性,提高数据库管理系统的执行效率,可以考虑使用视图、存储过程及表的触发器来实现某些功能。本设计可考虑如下数据库对象:

(1) 指定进行信息查询时,可以设计一些存储过程,如以学生学号或工作证号为输入参数返回指定学生或教师的基本信息。

(2) 统计每位教师的授课数和学时数。

(3) 浏览供选课程表。设计一个视图,返回可供选课的课程的基本信息及授课老师信息。

(4) 为提高检索性能,为表创建索引。

(5) 为新入学/休学/退学创建 INSERT、DELETE、UPDATE 触发器,实现学院各系学生人数的自动更新。

6.2.3 图书管理系统

图书管理系统其实是一个很复杂的信息管理系统,它包括很多分类、检索等方面的内容。因为其复杂性,建立这样一个系统更加能体现出运用 SQL Server 数据库进行数据处理的优越性。本课题将实现一个简化的图书管理系统的功能。

6.2.3.1 系统功能设计

(1) 信息录入功能

① 添加新图书信息。当图书馆收藏新图书时,系统向用户提供新图书信息录入功能,由于同一种书可能会有多本,因此新图书的信息有两类:某一个 ISBN 类别的图书信息,包括:ISBN 书号、图书类别、书名、作者、出版社、出版日期、价格、馆藏数量、可借数量、图书简介;另一个具体到每一本书的信息,包括:ISBN 书号、图书书号、是否可借。每一个 ISBN 书号和同一个 ISBN 书号的多本书之间是一对多的关系,每一本书的图书书号是唯一的。

② 添加读者信息,用于登记新读者信息。包括:借书证号、姓名、性别、出生年月、身份证号、职称、可借数量、已借数量、工作部门、家庭住址、联系电话等。

③ 借阅信息,用于登记读者的借阅情况信息。包括:借书证号、借阅书号、借出日期、归还日期等信息。归还日期为空值表示该图书未归还。

(2) 数据修改和删除功能

① 修改和删除图书信息。图书被借出时,系统需要更新图书的可借数量,当可借数量为 0 时,表示该图书都已被借出。当输入的图书信息有错误或需要进行更新时,可修改图书信息;当一种图书所有馆藏图书已损毁或遗失并且不能重新买到时,该图书信息删除。

② 修改和删除读者信息。当读者的自身信息发生变动,如部门间调动或调离本单位,或违反图书馆规定需要限制其可借阅图书数量时,需要修改读者信息。

③ 还书处理。读者归还图书时,更新图书借阅信息表中的归还日期,读者信息表中的已借数量有 ISBN 类别信息表中该图书的可借数量。

(3) 查询和统计功能

① 图书查询功能。根据图书的各种已知条件来查询图书的详细信息,对书名、作者、出版社、ISBN 书号等支持模糊查询。

② 读者信息查询。输入读者的借书证号、姓名、工作部门等信息,查询读者的基本信息。对查询到的每一个读者,能够显示其未归还的图书编号和书名。

③ 查询所有到期未归还的图书信息。要求结果显示图书编号、书名、读者姓名、借书证号、借出日期等信息。

④ 打印、统计指定读者一段时间内的某类图书或所有类别图书借阅次数及借阅总次数。

6.2.3.2 数据库完整性设计

设计者应认真分析和思考各个表之间的关系,合理设计和实施数据完整性原则。

(1) 给每个表实施主键及外键约束。

(2) 设定缺省约束,如性别。

(3) 设置非空约束,如图书信息表中的书名。

(4) 实施 CHECK 约束,如 ISBN 类别表中的可借数量小于馆藏数量。

(5) 实施规则,如身份证号码必须为 15 位或者 18 位。

6.2.3.3　SQL Server 数据库对象设计

(1) 设计一个存储过程,以图书编号为输入参数,返回借阅该图书但未归还的读者姓名和借书证号。

(2) 读者资料查询:设计一个有多个输入参数的存储过程,返回读者的详细信息。

(3) 到期图书查询:设计一个视图,返回所有逾期未归还图书的编号、书名、读者姓名等信息。

(4) 统计图书借阅次数:设计一个以两个日期作为输入参数的存储过程,计算这一段时间内各类别图书被借阅的次数,返回图书类别、借阅次数的信息。

(5) 加快数据检索速度,用图书编号为图书信息表建立索引。

(6) 为读者信息表创建一个删除触发器,当一个读者调出本单位时,将此读者的资料从读者信息表中删除。注意实施业务规则:有借阅书的读者不得从读者信息表中删除。

(7) 借阅处理:为借阅表设计 INSERT 触发器,在读者借阅时更改 ISBN 类别信息表,且可借数量减 1,图书信息表是否可借列的值变为"不可借",读者信息表中该读者已借阅数加 1。

(8) 还书处理:为借阅信息表设计 UPDATE 触发器,在读者借阅时更改 ISBN 类别信息表,且可借数量减 1,图书信息表是否可借列的值变为"不可借",读者信息表中该读者已借阅数加 1。

6.2.4　客房管理系统

在当今经济和商务交往日益频繁的状况下,宾馆服务行业正面临客流量骤增的压力,须借助计算机信息技术对宾馆服务进行管理。客房管理系统可以说是整个宾馆计算机信息管理系统的中心子系统,因为宾馆最主要的功能就是为旅客提供客房。

设计客房管理这样一个系统,可以涉及到大多数 SQL Server 数据库的重要数据库对象、重要功能和特性。通过这个课程设计可以加深对这些 SQL Server 数据库知识的学习、理解,积累在实际工程应用中运用各种数据库对象的经验,使学生掌握使用应用软件工程开发工具开发数据库管理系统的基本方法。

6.2.4.1　系统功能设计

系统功能是在实际开发设计过程中经过调研、分析用户需求,和用户一起确定下来,是系统为满足用户需求所应完成的功能。本课程设计模拟一个小型客房管理系统。本系统要求实现以下主要功能:

(1) 数据录入功能

在本系统中提供客人信息登记功能。可以录入客人的姓名、性别、年龄、身份证号码、家庭住址、工作单位、来自地的地名、入住时间、预计入住天数、客房类别、客房号、离店时间以及缴纳押金金额信息。在客人退房时,系统根据输入的离店时间及客房单价自动计算客人住宿费金额。

(2) 数据查询功能

① 查某类客房的入住情况及空房情况,显示所有该类客房空房数目和客房号。

② 根据客人姓名、来自地的地名、工作单位或家庭等信息查询客人信息;根据客房号查询入住客人的信息。

③ 查询某个客人住宿费用情况,显示客人缴纳押金金额、实际入住天数、客房价格、实际住宿费、住宿费差额及余额等信息。

④ 查询所有入住时间达到或超过预计入住天数的客人。

(3) 数据统计功能

① 统计一段时间内各类客房的入住情况。

② 统计全年各月份的客房收入。

③ 统计一段时间内各类客房的入住率。

6.2.4.2 数据完整性设计

设计者应认真分析和思考各个表之间的关系,合理设计和实施数据完整性原则。

(1) 给每个表实施主键及外键约束。

(2) 设定缺省约束。如性别,入住时间用 GETDATE();如三个表中所有货币类型的列都定义为缺省值。

(3) 设置非空约束。如客房类型名。

(4) 实施 CHECK 约束。如所有客人的离店时间都不可能小于入住时间。

(5) 实施规则。如身份证号码必须为 15 位或者 18 位。

6.2.4.3 SQL Server 数据库对象设计

系统需要设计并创建的视图、触发器和存储过程如下:

(1) 客人选择客房处理

设计一个存储过程,实现当客人来到饭店时,客人将告知饭店服务员自己所需的客房类型,服务员在系统中选择指定的客房类型后系统将显示所有空余的该类型客房,并显示该类型客房价格供客人参考选择。

(2) 客人入住登记处理

设计一个触发器,实现客人入住登记操作完成后,入住的客房状态应该及时做相应的改变,并记录客人的序号供以后查询。

(3) 客人离店退房处理

设计一个触发器,实现客人在离店退房时,服务员输入客人的退房时间,然后要计算出客人的住宿费用,以便于客人结账。同时,系统应该将客人所退客房的状态更改为"空",以便再次接待下一位客人入住。

(4) 客人信息查询处理

设计一个带输入参数的存储过程,服务员按客人的部分资料查询客人的全部信息以及客人住在哪一间客房。

(5) 查询住宿时间到期的客人

设计一视图,使得客房管理服务员可根据客人入住时登记的预住天数收取相应押金,当客人住宿时间达到预住天数时就应该通知客人,以便客人补交押金或退房。

(6) 统计一段时间内各类客房的入住率。

设计一个存储过程,实现饭店经营决策者能在一段时间内以各类型客房的入住情况为依据,获得入住率,从而调整饭店的经营策略。

入住率的计算公式:入住率=该类客房时间段内有客人的天数之和/时间段总天数/该类客房数。

6.2.5 民航订票管理系统

民航订票管理系统主要是为机场、航空公司和客户三方服务。航空公司提供航线和飞机的资料,机场则对在本机场起飞和降落的航班和机票进行管理,而客户能得到的服务应该有查询航班路线和剩余票数,以及网上订票等功能。

客户可以分为两类:一类是普通客户,对于普通客户只有普通的查询功能和订票功能,没有相应的票价优惠;另一类是经常旅客,需要办理注册手续,但增加了里程积分功能和积分优惠政策。机场还要有紧急应对措施,在航班出现延误时,要发送相应的信息。

6.2.5.1 系统功能设计

系统功能是在实际开发设计过程中经过调研、分析用户需求,和用户一起共同确定下来,是系统为满足用户需求所应完成的功能。本课程设计模拟一个小型民航定票管理系统。

航空客运订票的业务活动包括:查询航线、客票预订和办理退票等。试设计一个航空客运订票系统,以使上述业务可以借助计算机来完成。

(1) 每条航线所涉及的信息有:终点站名、航班名、飞机号、飞行周日(星期几)、乘员定额、余票量、已订票的客户名单(包括姓名、订票量、舱位等级 1,2 或 3)以及等候替补的客户名单(包括姓名、所需票量)。

(2) 作为示意系统,全部数据可以只放在内存中。

(3) 系统能实现的操作和功能如下:

① 查询航线:根据旅客提出的终点站名输出下列信息:航班号、飞机号、星期几飞行,最近一天航班的日期和余票额。

② 承办订票业务:根据客户提出的要求(航班号、订票数额)查询该航班票额情况,若尚有余票,则为客户办理订票手续,输出座位号;若已满员或余票额少于订票额,则需重新询问客户要求。若需要,可登记排队候补。

③ 承办退票业务：根据客户提供的情况（日期、航班），为客户办理退票手续，然后查询该航班是否有人排队候补，首先询问排在第一的客户，若所退票额能满足客户的要求，则为客户办理订票手续，否则依次询问其他排队候补的客户。

6.2.5.2 数据完整性设计

设计者应认真分析和思考各个表之间的关系，合理设计和实施数据完整性原则。
(1) 给每个表实施主键及外键约束。
(2) 设定缺省约束。如性别，如有二个表中货币类型的列都将其定义为缺省值。
(3) 设置非空约束。如客户姓名。
(4) 实施 CHECK 约束。如飞机的座位数不能小于 0。
(5) 实施规则。如编号及飞机型号的一些设置规定。

6.2.5.3 SQL Server 数据库对象设计

系统需要设计并创建的视图、触发器和存储过程如下：
(1) 客户订票过程

设计一个存储过程，实现客人订票处理。

提示：需先查看客户是否为特殊客户，若不是，则票价不打折扣；否则如果客户累计航程超过 10 万公里，票价打九折；超过 20 万公里，打八折。获得确切票价后加入客户订票信息表中，并将客户新订票里程的信息累计到用户信息表中。注：须查看客户订票后，是否超过可容纳的座位数目，如果超过，取消所有操作。

(2) 飞机信息更新处理

设计一个触发器，实现若在飞机信息表中删除了一架飞机，则同时删除该航班的信息。

(3) 航班延误处理

设计一个触发器，实现若航班产生延误，则发出信息通知客户。

6.2.6 学生信息管理系统

学生信息管理系统是对学生的基本信息和成绩信息进行管理，主要包括添加、修改和删除学生的基本信息及课程的基本信息；录入、修改和删除学生的成绩信息，对基本信息、成绩信息进行查询、排序及统计等操作，从而实现学生信息管理的自动化与计算机化。本课题将实现一个简化的学生信息管理系统。

6.2.6.1 系统功能模块设计

(1) 学生基本信息管理模块：对学生的基本信息（学号、姓名、出生年月、性别、班级、所在系等）进行综合管理，可以添加、修改及删除学生的基本信息（可在同一界面完成该模块的功能，也可以分多个界面来完成）。

(2) 成绩管理模块：对学生所选课程的成绩信息（学号、课程名、成绩等）进行综合管理，可以添加、修改及删除基本信息（可在同一界面完成该模块的功能，也可以分多个界面来完成）。

（3）课程信息管理模块：对课程信息（课程号、课程名、学分、任课教师等）进行综合管理，可以添加、修改及删除课程的基本信息（可在同一界面完成该模块的功能，也可以分多个界面来完成）。

（4）查询模块：

① 学生基本信息的查询：根据学生的已知条件来查询学生的详细信息，对姓名、学号、班级、系名等支持模糊查询。

② 课程基本信息的查询：根据课程的信息来查询课程的详细信息。

③ 查询学生的选课情况、查询学生所选课程的成绩。

（5）统计模块：根据不同课程对学生成绩进行统计，求平均分、总分等；根据不同的分数区间进行人数统计等。

6.2.6.2 数据库完整性设计

设计者应认真分析和思考各个表之间的关系，合理设计和实施数据完整性原则。

（1）给每个表实施主键及外键约束。

（2）设定缺省约束，如性别。

（3）设置非空约束，如学生的姓名。

（4）实施 CHECK 约束，如性别只能为'男'或'女'两值。

（5）实施规则，如学号必须为 8 位数字，并前四位为 2004 等。

6.2.6.3 SQL Server 数据库对象设计

（1）设计一个存储过程，以学号为输入参数，返回该学生未选修的课程号和课程名。

（2）查询模块中：可自行设计一个有多个输入参数的存储过程，并返回详细信息。

（3）在查询模块中，自行设计一个视图，如返回所有选课学生的学号、姓名、课程号、课程名信息。

（4）统计某门课程的平均分：设计一个以课程名作为输入参数的存储过程，计算选修了该门课程学生的平均成绩（成绩为空的学生不参与统计），返回课号、课程名、平均分信息。

（5）加快数据检索速度，用姓名为学生信息表建立索引。

（6）为学生信息表创建一个删除触发器，当一学生被删除时，需将此学生的成绩资料从成绩信息表中删除。

（7）为课程表设计 update 触发器，在课程表中当更改了课程号，相应的在成绩表中同一门课的课号也应改变。

6.2.7 长途汽车信息管理系统

6.2.7.1 系统功能设计

系统功能是在实际开发设计过程中经过调研、分析用户需求，和用户一起共同确定下来，是系统为满足用户需求所应完成的功能。本课程设计模拟一个长途汽车信息管理系统。

长途汽车信息管理系统的业务活动包括：查询线路信息、汽车信息、票价信息，完成客票预订和办理退票等。

(1) 线路信息管理模块：对线路的基本信息（出发地、目的地、出发时间、所需时间等等）进行综合管理，可以添加、修改及删除线路的基本信息。

(2) 汽车信息管理模块：对汽车信息（汽车的类型、最大载客量等）进行综合管理，可以添加、修改及删除汽车信息。

(3) 票价信息管理模块：对票价信息（线路、汽车类型、票价等）进行综合管理，可以添加、修改及删除票价的基本信息。

(4) 查询模块：

① 线路信息的查询：根据线路的已知条件（如目的地和出发时间）来查询线路的详细信息，需要支持模糊查询。

② 汽车基本信息的查询：根据汽车的某个或某些信息来查询汽车的详细信息。

③ 票价基本信息的查询：查询某个线路所有票价。

④ 统计模块：统计某段线路的最低、最高票价并列出相应的汽车详细信息。

6.2.7.2 数据完整性设计

设计者应认真分析和思考各个表之间的关系，合理设计和实施数据完整性原则。

(1) 给每个表实施主键及外键约束。

(2) 设定缺省约束，如最大载客量。

(3) 设置非空约束，如票价。

6.2.7.3 SQL Server 数据库对象设计

系统需要设计一个触发器，实现在汽车信息表中若删除了某种类型的汽车，同时需删除该类汽车的票价信息。

6.2.8 其他参考选题

- 高校科研信息管理系统
- 高校人事管理系统
- 学校物资管理系统
- 汽车租赁管理系统
- 新闻发布系统
- 餐厅管理系统
- 宾馆管理系统
- 物业管理系统
- 小商店销售管理系统
- 库房入库出库管理系统

- 教室管理系统
- 通讯录系统

6.3 课程设计报告要求

设计报告内容参照以下提纲书写:
(1) 摘要。
(2) 需求分析。
(3) 数据库概念结构设计。
(4) 数据库逻辑结构设计。
(5) 数据流图及系统功能模块图。
(6) 程序源代码及其说明。
(7) 总结:总结课程设计的过程、体会及建议。
(8) 参考文献。

设计报告排版参考格式：

《数据库原理及应用》课程设计

设计报告

院系_____ 班级_____

学号_____ 姓名_____

指导教师 _____

题　目　_____

　　　　_____

年　　月　　日

课程设计报告打印格式

报告组成:
1. 封面;2. 摘要;3. 目录;4. 正文;5. 参考文献。
注:封面不加页码,摘要为第一页,每一章需分页。
排版要求:
(页面设置:纸张为 A4,左 3cm,右 1.5cm,其他默认)
(页眉设置:以小五号宋体键入"课程设计",居中)
(页脚设置:插入页码,居中)
摘　要(小四号宋体加粗):内容(小四号宋体)
目　录(三号黑体加粗,居中)
1　学生成绩管理
1.1　需求分析
1.1.1　用户需求
(四号黑体,注明各章节起始页码,标题与页码用"……"相连)
(正文内容一律用小四号宋体,行距为固定值 22 磅)
(标题格式如下:)
1　学生成绩管理
1.1　需求分析
1.1.1　用户需求
(四号宋体加粗,段前、段后均为 6 磅)
参考文献:(四号宋体加粗,居中)
[1]　×××.中国高等教育.北京:中国教育报刊社,2006:17-18.
　　(五号宋体:作者.书名[M].出版地:出版社,时间:页码.)
[2]　×××,×××等.×××研究[J].××学报,2005,17(8):124-129.
　　(五号宋体:作者.论文名[J].刊名,年,卷(期):页码.)
[3]　Maham G D. Mang-Chemistry[M]. New York:Plenum,1989.
　　(同[1])
[4]　Curzon F,＊＊＊,＊＊＊,ect..Efficiency of…[J]. Am. J. Phys.,1975,4(1):22-25.
　　(同[2])

参考文献

[1] 王珊,萨师煊. 数据库系统概论(第四版). 北京:高等教育出版社,2006.

[2] 赵斌. SQL Server 2008 应用开发案例解析. 北京:科学出版社,2009.

[3] 李爱武. SQL Server 2008 数据库技术内幕. 北京:中国铁道出版社,2012.

[4] 郝安林等. SQL Server 2008 基础教程与实验指导. 北京:清华大学出版社,2012.

[5] (美)Robert Vieira 著 杨华,腾灵灵译. SQL Server 2008 高级程序设计. 北京:清华大学出版社,2010.

[6] 尹相志等. SQL Server 2008R2 Reporting Services 报表服务. 北京:中国水利水电出版社,2012.